Which Way Did the Bicycle Go?

...And Other Intriguing Mathematical Mysteries

Joseph D. E. Konhauser

Dan Velleman

Stan Wagon

MATHEMATICAL ASSOCIATION OF AMERICA

DOLCIANI MATHEMATICAL EXPOSITIONS – No.18

THE
DOLCIANI MATHEMATICAL EXPOSITIONS

Published by
THE MATHEMATICAL ASSOCIATION OF AMERICA

Which Way Did the Bicycle Go?

...and Other Intriguing Mathematical Mysteries

Library of Congress Catalog Card Number 95-81495

Complete Set ISBN 0-88385-300-0
Vol. 13 ISBN 0-88385-325-6

Printed in the United States of America

Current printing (last digit):
10 9 8 7 6 5 4 3 2 1

The DOLCIANI MATHEMATICAL EXPOSITIONS series of the Mathematical Association of America was established through a generous gift to the Association from Mary P. Dolciani, Professor of Mathematics at Hunter College of the City University of New York. In making the gift, Professor Dolciani, herself an exceptionally talented and successful expositor of mathematics, had the purpose of furthering the ideal of excellence in mathematical exposition.

The Association, for its part, was delighted to accept the gracious gesture initiating the revolving fund for this series from one who has served the Association with distinction, both as a member of the Committee on Publications and as a member of the Board of Governors. It was with genuine pleasure that the Board chose to name the series in her honor.

The books in the series are selected for their lucid expository style and stimulating mathematical content. Typically, they contain an ample supply of exercises, many with accompanying solutions. They are intended to be sufficiently elementary for the undergraduate and even the mathematically inclined high-school student to understand and enjoy, but also to be interesting and sometimes challenging to the more advanced mathematician.

---

1. *Mathematical Gems,* Ross Honsberger
2. *Mathematical Gems II,* Ross Honsberger
3. *Mathematical Morsels,* Ross Honsberger
4. *Mathematical Plums,* Ross Honsberger (ed.)
5. *Great Moments in Mathematics (Before 1650),* Howard Eves
6. *Maxima and Minima without Calculus,* Ivan Niven
7. *Great Moments in Mathematics (After 1650),* Howard Eves
8. *Map Coloring, Polyhedra, and the Four-Color Problem,* David Barnette
9. *Mathematical Gems III,* Ross Honsberger
10. *More Mathematical Morsels,* Ross Honsberger
11. *Old and New Unsolved Problems in Plane Geometry and Number Theory,* Victor Klee and Stan Wagon
12. *Problems for Mathematicians, Young and Old,* Paul R. Halmos
13. *Excursions in Calculus: An Interplay of the Continuous and the Discrete,* Robert M. Young
14. *The Wohascum County problem Book,* George T. Gilbert, Mark Krusemeyer, and Loren C. Larson
15. *Lion Hunting and Other Mathematical Pursuits: A Collection of Mathematics, Verse, and Stories by Ralph P. Boas, Jr.,* edited by Gerald L. Alexanderson and Dale H. Mugler
16. *Linear Algebra Problem Book,* Paul R. Halmos
17. *From Erdős to Kiev: Problems of Olympiad Caliber,* Ross Honsberger
18. *Which Way Did the Bicycle Go? ...and Other Intriguing Mathematical Mysteries,* Joseph D. E. Konhauser, Dan Velleman, and Stan Wagon

Preface

> While realizing that the solution of problems is one of the lowest forms of mathematical research, and that, in general, it has no scientific value, yet its educational value cannot be overestimated. It is the ladder by which the mind ascends into the higher fields of original research and investigation. Many dormant minds have been aroused into activity through the mastery of a single problem.
>
> — Benjamin Finkel and John M. Colaw,
> founders of the *American Mathematical Monthly,* 1894
> (*Amer. Math. Monthly,* 1894, volume 1, number 1, page 1)

Joe Konhauser (1924–1992) was a great believer in the value of problem-solving activity. For 25 years he posted a Problem of the Week at Macalester College, and this book consists of 190 problems chosen from Joe's and, later, Stan Wagon's, posted problems. In order to encourage student participation, Joe developed a definite style in the problems he posted. They had to involve almost no prerequisites and be succinctly stated and inherently attractive. In short, his plan was to hook students immediately into thinking about the problem.

This book is aimed at everyone with an interest in problems, but in particular at teachers who want a source of problems for their students. We have taken what we believe to be the most attractive problems from the Konhauser collection. Because his files contain many student solutions and attempts at solutions, we are confident that they will appeal to students at the advanced high-school or beginning college level.

We have followed the traditional problem book layout. Part I contains the problems in no particular order except for the chapter and section groupings by field. Part II then contains the solution, historical and other notes, and often some auxiliary problems (without solution). While nothing could be more obvious than that a problem is more appreciated if one tries it without looking at the solution, there is something to be said for using the solution as a guide to whether one wishes to study the problem in more detail. The choice is yours. But many of these problems have surprising little twists in them and for the majority of them the solutions are quite short; thus even if you choose not to solve a problem yourself, we encourage you to at least take a guess at the answer before turning to the solution.

We find the computer to be very valuable in analyzing certain types of problems, and there are several in this collection for which a computer is essential for the solution. A good software package is just one more tool in the problemist's arsenal, and we encourage its use. In several of the solutions we have included a few lines of *Mathematica* code so that the reader can see what is involved in generating and analyzing computer output.

We have included reference information in all cases for which we knew it. Posing a good problem is harder than solving it, and we would like to fill in any missing attributions, so do let us know if you have information of this sort.

Finally, we must acknowledge our gigantic and, sadly, unpayable debt to Joe Konhauser. In his hands, the Problem of the Week was an opportunity to share with students the many surprising twists and patterns that can be found even in very elementary mathematics. And the students responded over the years with humor, perseverance, and occasional brilliance. Many of the solutions in this book come from the yellowed sheets of student submissions in the Konhauser files, and we therefore express our thanks to several generations of Macalester students. Faculty also contributed mightily to the solution files, and we gratefully acknowledge the many fine contributions of three Macalester professors: John Schue, Emil Slowinski, and the late John Howe Scott. The last two are chemists and their indefatigable efforts over the years show once again the universal appeal of attractive puzzles. We are also indebted to those who have provided us with problems for inclusion; we would like to especially mention Lee Sallows, an electrical engineer from Nijmegen, the Netherlands, who has invented many beautiful problems, three of which are included in this collection. We thank the many colleagues who have commented on our draft manuscript, providing many references or superior solutions: these include Frank Bernhart, Larry Carter, Hung Dinh, Woody Dudley, John Duncan, Curtis Greene, Jim Guilford, John Guilford, Richard Guy, John Hamilton, Joan Hutchinson, Murray Klamkin, Loren Larson, Jim Mauldon, Hugh Montgomery, Bruce Palka, and Stanley Rabinowitz. And we are grateful to Bev Ruedi at the MAA for her superb job of typesetting and layout.

Dan Velleman
Amherst College, Amherst, Mass.
djvelleman@amherst.edu

Stan Wagon
Macalester College, St. Paul, Minn.
wagon@macalester.edu

P.S. Stan Wagon maintains the Problem of the Week tradition at Macalester College and sends out the weekly problems by e-mail; contact him if you would like to be added to the mailing list.

Contents

CHAPTER 1

Plane Geometry

1.1 Locus

1. Which Way Did the Bicycle Go?

> "This track, as you perceive, was made by a rider who was going from the direction of the school."
> "Or towards it?"
> "No, no, my dear Watson . . . It was undoubtedly heading away from the school."
> —Sherlock Holmes, during his visit to the Priory School

Here's a mystery that is truly worthy of Sherlock Holmes! Imagine a 20-foot wide mud patch through which a bicycle has just passed, with its front and rear tires leaving tracks as illustrated. In which direction was the bicyclist travelling?

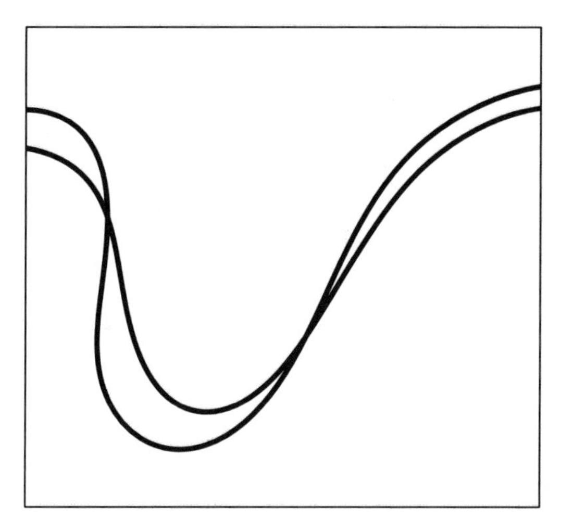

2. Where Can the Third Vertex Live?

Given $\triangle ABC$, describe the set of points X for which there exists a point D on BC such that $\triangle ADX$ is equilateral.

3. Seeing a 45° Angle

Given a square, determine the set of points P in the plane of the square for which the square, viewed from P, presents a 45° angle.

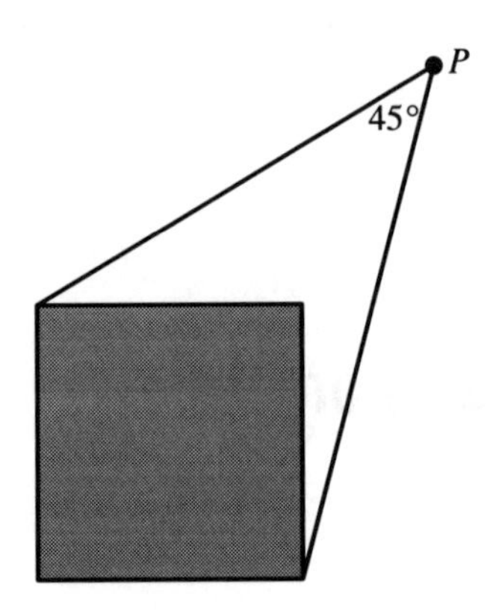

4. A Narrow Path

In the diagram, ABC represents an equilateral triangle (whose size is not fixed) that straddles a 120° angle as illustrated. Imagine that this triangle moves so that the vertices B and C move along the lines VU and VW, respectively (and the size of the triangle changes accordingly). What is the shape of the path traced out by vertex A during such movement?

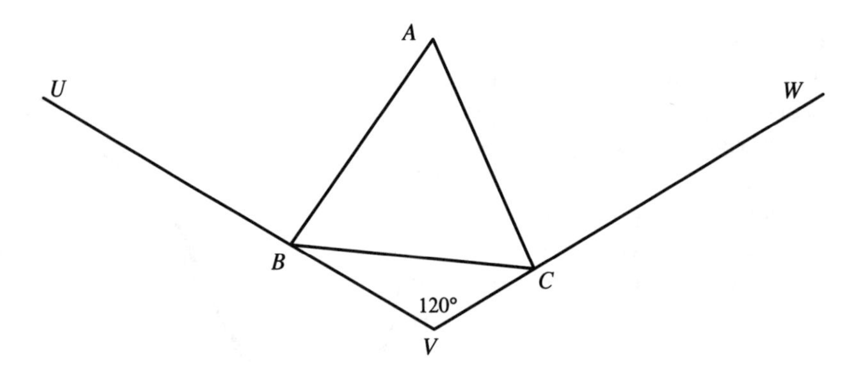

5. Test Your Intuition

Let S be a semicircle rising from the origin and lying on the positive x-axis, as illustrated. Consider also a circle C centered at the origin and let A and B be the points of intersection of C with the positive y-axis and with S, respectively. Extend the line AB rightward, letting X be its intersection with the x-axis. What happens to X as C becomes smaller and smaller, its radius approaching zero?

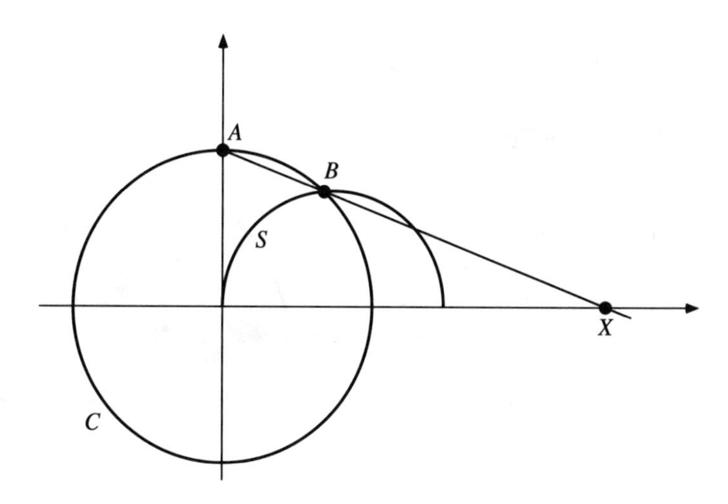

6. Disks on a Circle

Let P be a fixed point on a circle C. What is the shape of the region covered by the circular disks with centers on C and with boundaries passing through P?

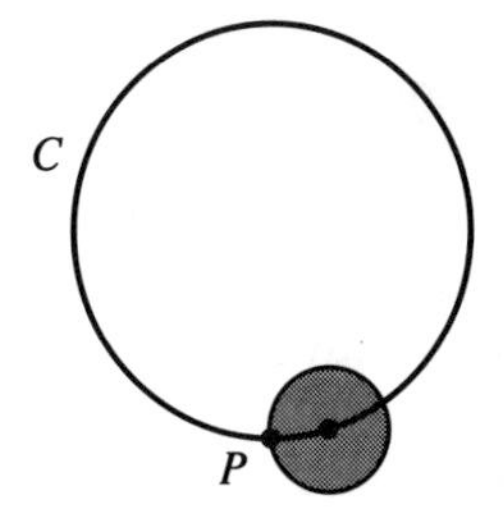

7. Don't Cut Corners—Fold Them

Suppose the first quadrant of the x-y plane is a giant sheet of paper. Fix a constant K and imagine that the corner at $(0, 0)$ is folded over onto a point P on the sheet in such a way that the triangle folded over has area K. Describe the set of points that can occur as P.

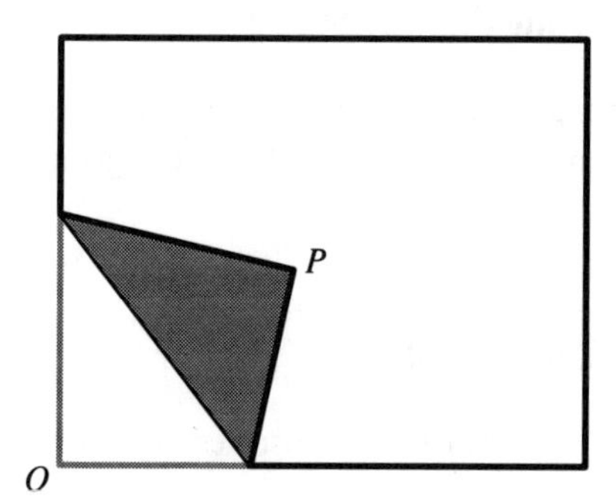

1.2 Dissection

8. Straighten these Curves

The diagram shows a circle of radius 1, with the boundary of the shaded portion consisting of three circular arcs of radius 1 whose centers are equally spaced on the ambient circle. Dissect the unshaded portion of the circle's interior into pieces that can be reassembled to form a rectangle.

9. Cut the Triangle

Can you cut an arbitrary triangle into pieces so that the pieces can be rotated and translated (but not flipped) so as to form the mirror image of the given triangle? It can be done in just two cuts.

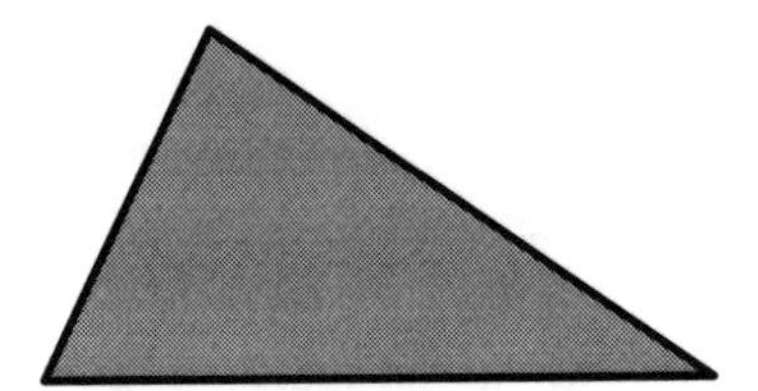

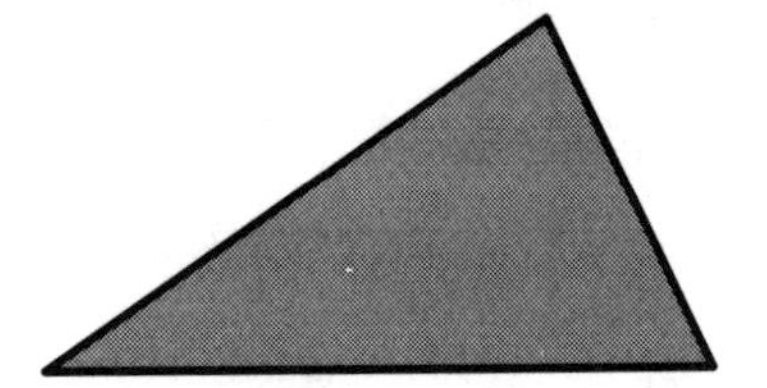

10. A Triangle Duplication Dissection

Show that any triangle can be dissected by straight cuts into four pieces that can be arranged to form two triangles similar to the given triangle.

11. A Square Triangulation

Can you cut a square into seven isosceles right triangles, no two of which are congruent?

12. Equilateral Into Isosceles

Dissect an equilateral triangle into five isosceles triangles so that

(a) none of the five isosceles triangles is equilateral;

(b) exactly one of the five isosceles triangles is equilateral;

(c) exactly two of the five isosceles triangles are equilateral.

13. Solitaire on a Chessboard

Start with a chessboard and place a marker on a square. Then duplicate that marker and place the new marker somewhere else on the board. Then duplicate that configuration of two markers and place the resulting configuration of two markers on empty squares (you may translate in any direction, not just parallel to the axes, but you may not rotate the configuration, and you must move the whole configuration at once). Then duplicate the resulting configuration of 4 markers and place the new set of 4 on free squares (if you can). It is easy, by a judicious choice of moves, to cover the entire 8×8 board in this way, as illustrated.

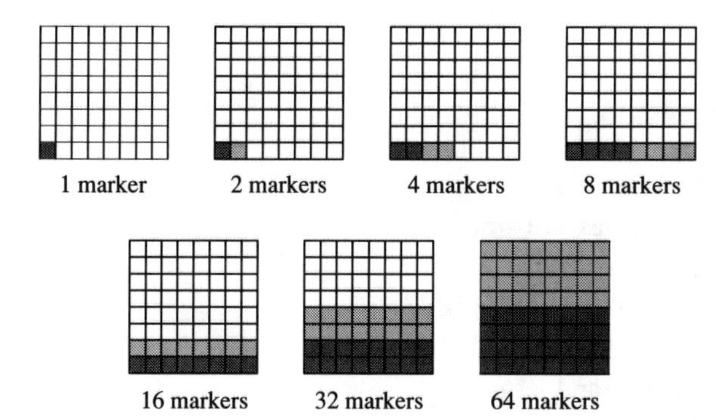

Now, suppose the board is 7×7. Then, because 49 is not a power of 2, the most squares you could possibly cover would be 2^5, or 32.

True or False: One can cover 32 squares of a 7×7 board by starting with a single square and repeatedly duplicating and translating the configuration of covered squares into open space.

14. A Notorious Tiling Problem

Experienced artisans tend to use the tools at hand. In the mid-1980s N. G. de Bruijn proved the result of this problem using complex numbers. In 1986, after Hugh

Montgomery publicized the problem, the mathematical community amused itself by coming up with fourteen different elementary proofs. Can you find one?

Show that whenever a rectangle is tiled with rectangles, each of which has at least one integer side, then the tiled rectangle has at least one integer side.

1.3 Triangles

15. An 80°-80°-20° Triangle

Suppose $\triangle ABC$ is isosceles with $AB = AC$ and $\angle BAC = 20°$. And suppose P is on side AB such that $AP = BC$. Determine $\angle ACP$.

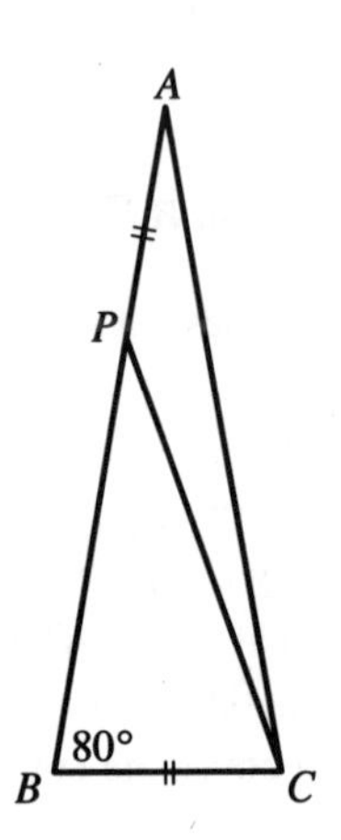

16. A Decomposition of Unity

A point P is chosen inside a triangle ABC, and lines are drawn through P parallel to the sides of the triangle. Let a, b, and c be the lengths of the sides BC, AC, and AB respectively, and let a', b', and c' be the lengths of the middle segments of these sides, as indicated in the figure. Show that

$$\frac{a'}{a} + \frac{b'}{b} + \frac{c'}{c} = 1.$$

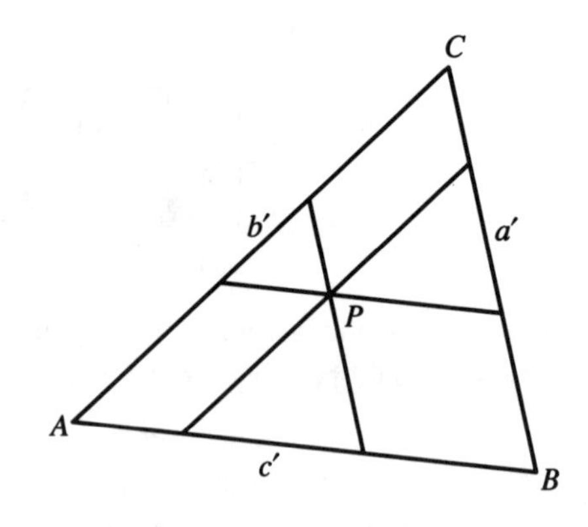

17. Find the Missing Altitude

Two of the altitudes of a triangle are 9 and 29. If the third altitude is also a positive integer, what values can it have?

18. An Isosceles Chain

For an isosceles triangle T with base angle B, let $K(T)$ be an isosceles triangle whose vertex angle is B. For example, if T is an 80°-80°-20° triangle, then $K(T)$ is a 50°-50°-80° triangle. Find (with proof) the base angles of the triangles in the longest chain of triangles $T, K(T), K(K(T)), \ldots$ such that T is not equilateral and such that all the angles in all the triangles in the chain have an integer number of degrees.

19. Nested Triangles

A circle is inscribed in a nonequilateral triangle and the points of tangency are taken as the vertices of a second triangle. In this second triangle a circle is inscribed, and its points of tangency form the vertices of a third triangle. If this process is continued to create an infinite sequence of nested triangles, will any two triangles in the sequence be similar?

20. Avoiding Equilaterals

Can the plane be divided into two disjoint subsets so that neither set contains the vertices of an equilateral triangle?

21. The Square on the Hypotenuse

An integer-sided square is inscribed in an integer-sided right triangle so that one side of the square lies on the hypotenuse. What is the smallest square that can arise in this way?

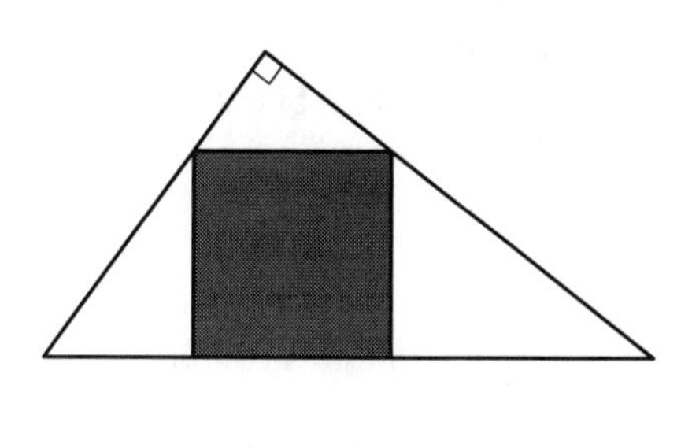

22. A Hexagon-Triangle Hinge

In the diagram, P is the center of a regular hexagon with a vertex C at which it touches an equilateral triangle with center Q. Let M be the midpoint of AB. Show that $\angle PMQ$ is a right angle.

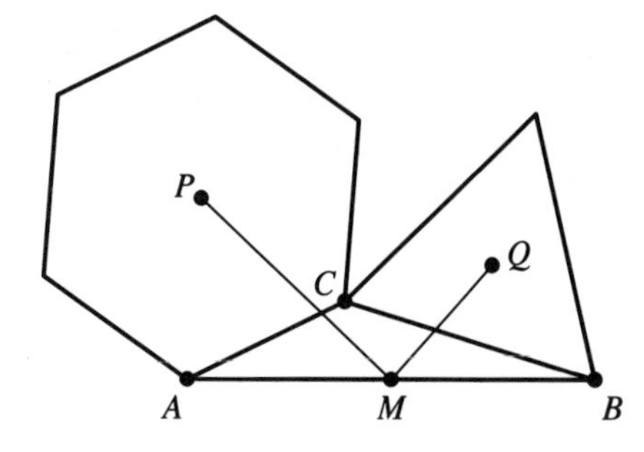

23. How Many Isosceles Triangles?

Given a segment AB in the plane and a line L in the same plane, let $n(L)$ be the number of points C on L such that $\triangle ABC$ is isosceles. For example, if L is the perpendicular bisector of AB, then $n(L)$ is infinite. What are the other possible values of $n(L)$?

24. A Triangle Bisection

In $\triangle ABC$ let M be the midpoint of side AB. Through M draw a line parallel to the angle bisector of $\angle ACB$, intersecting the triangle again in point N. Show that MN bisects the perimeter of $\triangle ABC$.

25. A Cut Through the Centroid

A line is drawn through the centroid of a triangle, cutting it into two pieces. Show that if the area of the triangle is 1, then the area of each piece must be at least $\frac{4}{9}$.

26. A Common Ratio

In the diagram, $\triangle ABC$ is equilateral, points M and N are the midpoints of sides AB and AC, respectively, and F is the intersection of line MN with the smaller arc AC of the circle circumscribing $\triangle ABC$. Show that $MF/MN = MN/NF$ and determine their common value.

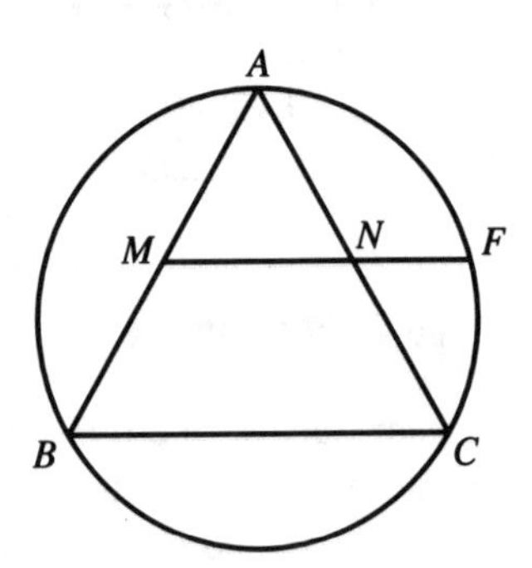

27. A 3-4-5 Triangle Problem

Triangle ABC is equilateral and P is in its interior. The distances PA, PB, PC are 3, 4, 5, respectively. What is the side-length of the triangle?

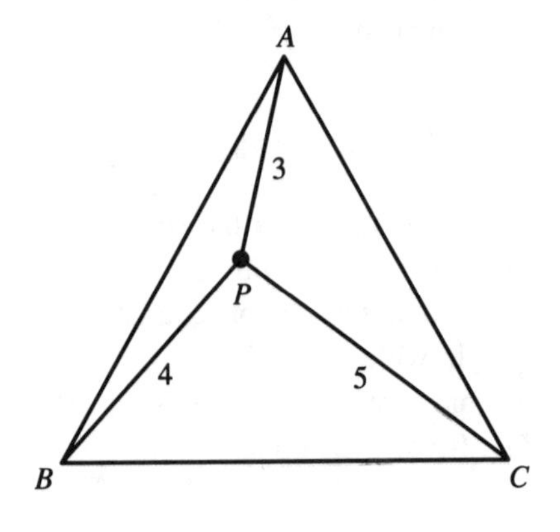

28. Triangles from a Triangle

Suppose ABC is a triangle in the plane. Find all points X such that the segments XA, XB, and XC can be translated—moved without change in length or direction—so as to form a triangle.

29. An Equilateral Triangle from a Circle and Hyperbola

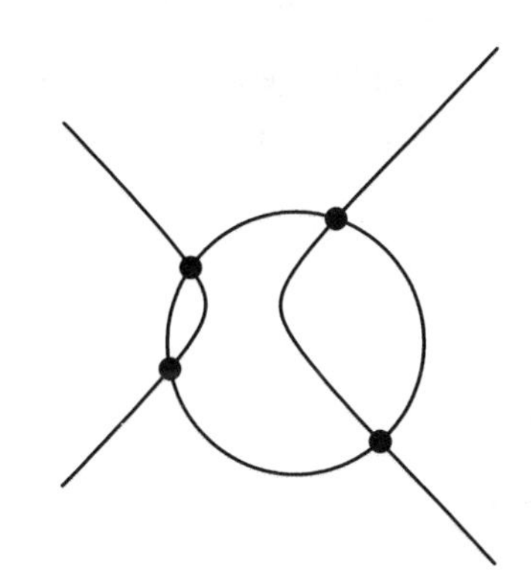

The hyperbola $x^2-y^2+ax+by=0$ and circle $x^2+y^2=a^2+b^2$ intersect in four points. Show that three of the four points form an equilateral triangle.

1.4 Circles

30. Abe Lincoln's Somersaults

Eight congruent disks (pennies, for example) are touching each other as in the diagram. If the dark disk is rolled, without slipping, around the other seven, how many complete somersaults will Abe Lincoln make as he rolls through one full circuit of the array?

31. Focus on This

A bicycle wheel rolls from left to right on a flat surface so that its center moves at a constant 15 mph (22 feet per second). Suppose a wheel of the bike is a solid disk covered with advertising in small print and a photographer with a stationary camera takes a picture of the wheel as it passes. Assume that a letter is not too blurry to read provided the speed of the letter when photographed is no greater than 11 feet per second. Which parts of the wheel will be legible in the photograph?

32. Circular Surprises

In the diagram, the large circle has radius one and the inscribed figures are a diameter, an equilateral triangle, and a square. Determine the radii of the shaded and unshaded circles.

33. Tangent Cuts

In the figure, A and B are the centers of the two circles. Lines through A tangent to the circle centered at B cut the circle centered at A at the points P and Q. Similarly, R and S are the intersection points of the circle centered at B with lines through B tangent to the circle centered at A. Show that the distance from P to Q is equal to the distance from R to S.

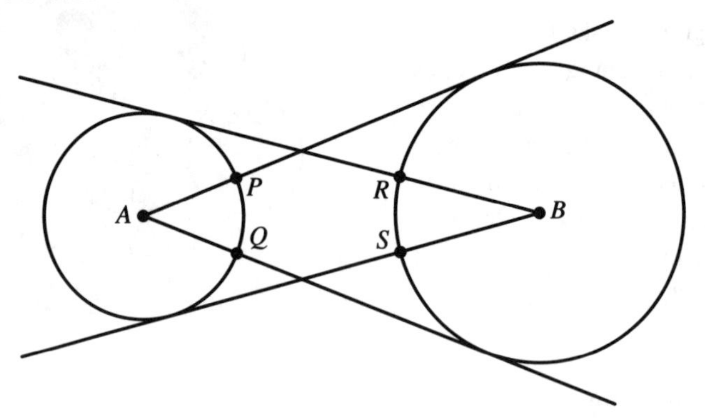

34. Five Circles

Five circles are tangent to two nonparallel lines and to each other as illustrated. The smallest radius is 4; the largest radius is 9. What is the radius of the middle circle?

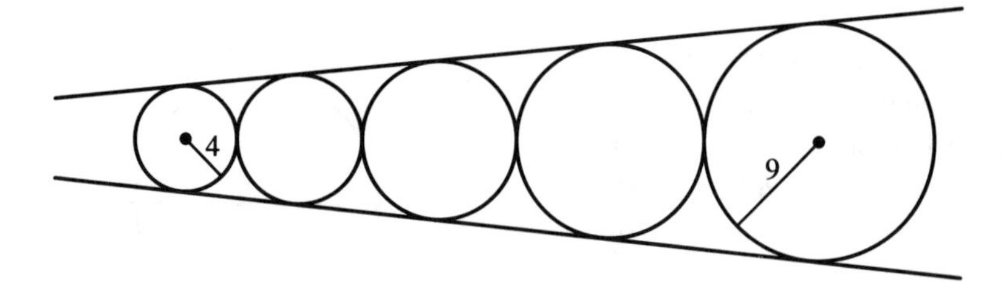

35. A Ring of Disks

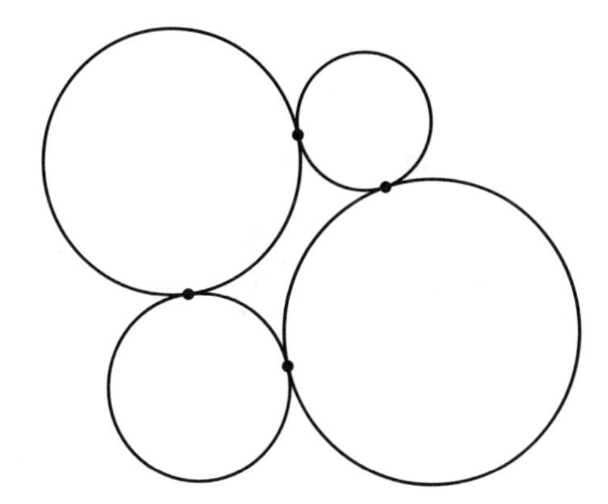

Four disks are arranged in a ring as shown. Prove that the four points of tangency lie on a circle.

36. A Circle Inside an Angle

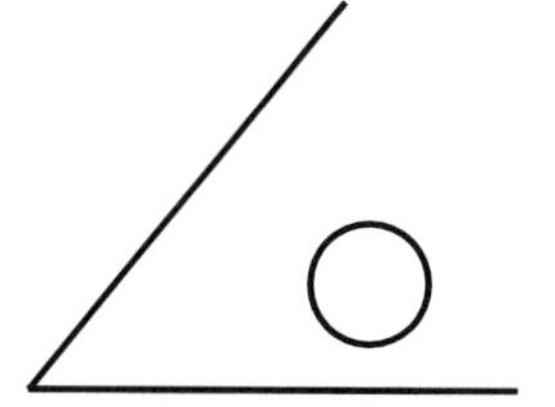

Given a circle inside an angle, as in the figure, for which point on the circle is the sum of the distances from that point to the sides of the angle least? For which point is it greatest?

37. A Ray that Pierces Concentric Circles

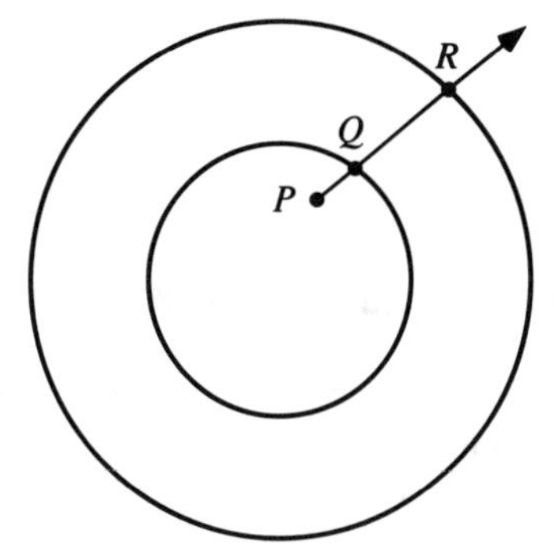

Suppose P is a point inside two concentric circles but different from their common center. A ray from P intersects the smaller circle in Q and the larger circle in R. For which direction is the length of line segment QR a maximum?

38. If You Lose Your Compass

Given a circle with center O, a line through O, and a point P not on the line or the circle, construct *with unmarked straightedge only* a line through P perpendicular to the given line.

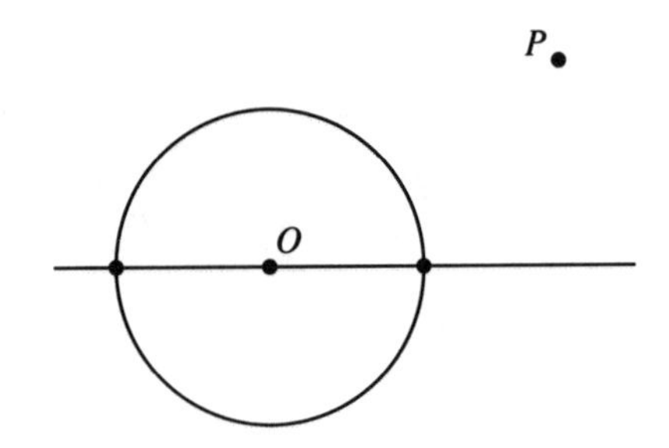

39. Circles in a Circle

In the figure, AB is a diameter of the large circle, and CD is perpendicular to AB. Circle 1 is tangent to all three sides of $\triangle ABC$, and circles 2 and 3 are tangent to AB, CD, and the large circle. Show that the radius of circle 1 is the average of the radii of circles 2 and 3.

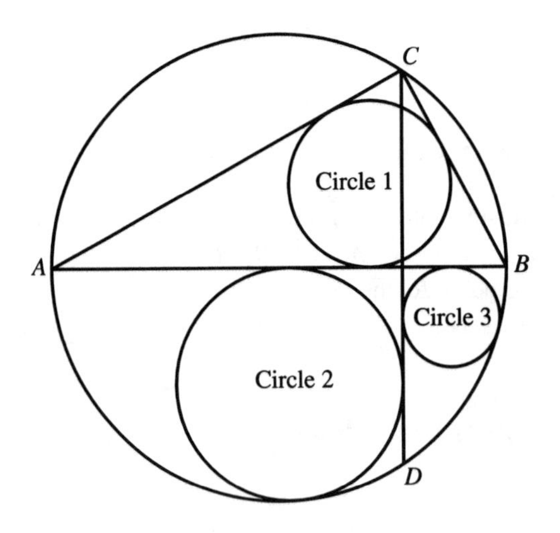

40. Get Close to the Circles

Let O_1, O_2, O_3, and O_4 be the centers of four nonoverlapping coplanar circles of unit radius. Show that if P is any point in the plane of the circles then $(PO_1)^2 + (PO_2)^2 + (PO_3)^2 + (PO_4)^2 \geq 6$.

41. An Ellipse and a Circle

Parallel lines are drawn tangent to an ellipse with semimajor axis a and semiminor axis b. A circle is then drawn tangent to the ellipse and tangent to both of the parallel lines. Show that the distance between the centers of the ellipse and the circle is $a + b$.

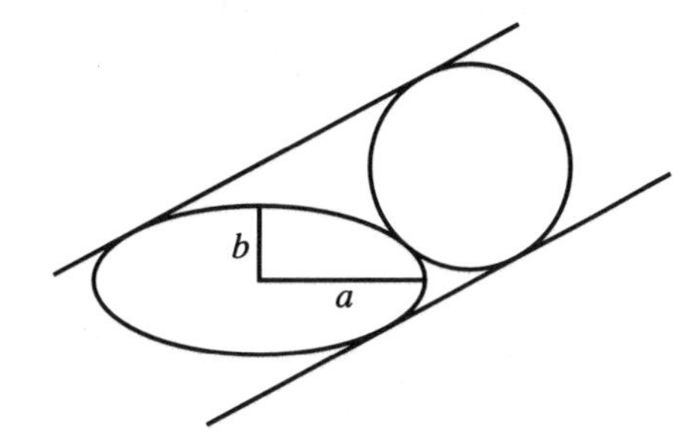

42. Three Circles in a Circle

In the diagram, the three small circles are congruent and mutually tangent and the large circle is tangent to all the small circles. The point P is on the large circle and PA, PB, and PC are tangents to the three small circles. Show that $PA + PB = PC$.

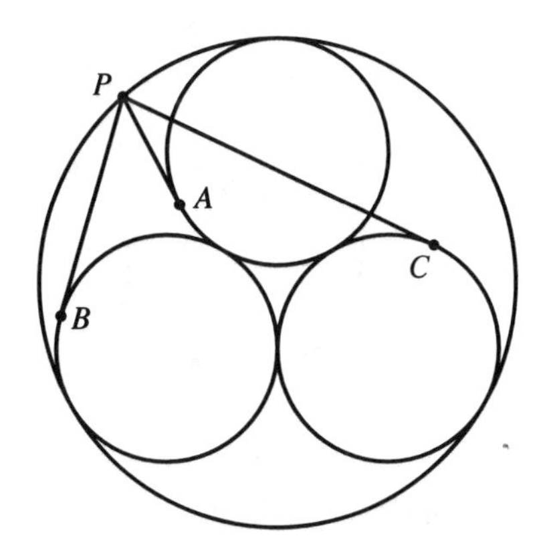

43. Two Nickels and Three Pennies

There are just two essentially different ways of arranging two nickels and three pennies in a ring so that each coin is tangent to two others and all five coins are externally tangent to a circle inside the ring (see color plate 1 following p. 16). For which arrangement is the radius of the inner circle larger?

44. Thirteen Bottles of Wine

All circles in this problem have radius 1. Place three circles, a, b, and c reading from left to right, in the first quadrant so that they lie on the x-axis, a is tangent to the y-axis, and the space between the centers of a and b (and the centers of b and c) is less than 4. Then place circles d and e above these so that d is tangent to a and b, and e is tangent to b and c. Then place three more circles, f, g, and h, above these so that g is tangent to d and e, while f is tangent to d and the y-axis, and h is tangent to e and the vertical line that forms the right-hand boundary to c.

Continue for two more rows, to get a top row of k, l, m. The configuration looks like 13 wine bottles in a rack, viewed end-on (see diagram). Prove that the top row is horizontal.

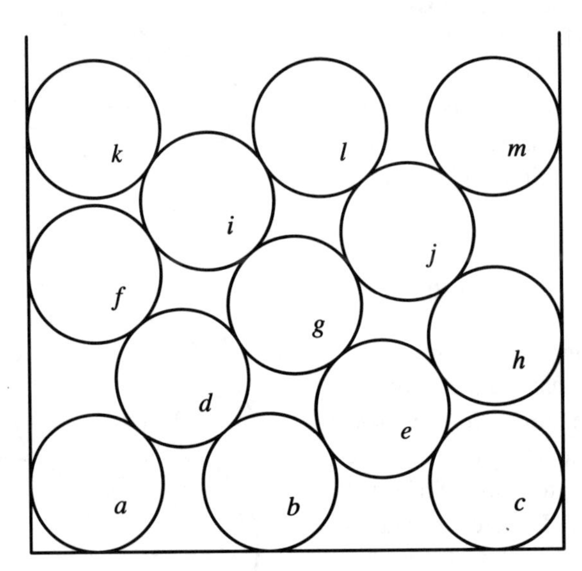

45. A Hexagon in a Circle

When a regular hexagon with side-length 1 is inscribed in a circle, the circle's radius is 1. Is there an integer $a > 1$ such that when a hexagon with sides, in order, 1, 1, 1, a, a, a, is inscribed in a circle, the circle has an integer radius?

1.5 Packing and Covering

46. A 6-Sided Peg in a Square Hole

Find the largest regular hexagon that fits into a square of side-length 2.

47. Packing 11 Squares

Without overlapping, 9 unit squares can be fit into a 3×3 square, but not into a smaller square. What is the smallest square into which you are able to fit 11 unit squares, again, without overlap?

48. Find the Rectangles

A regular 400-gon is tiled with nonoverlapping parallelograms. Prove that at least 100 of these parallelograms are rectangles.

49. Straightening Out a Circle

The diagram shows how to cover any convex quadrilateral and its interior with disjoint closed line segments, none having zero length.

(a) Is it possible to cover a triangle and its interior with disjoint closed line segments none having zero length?

(b) Can a circle and its interior be covered in such a way?

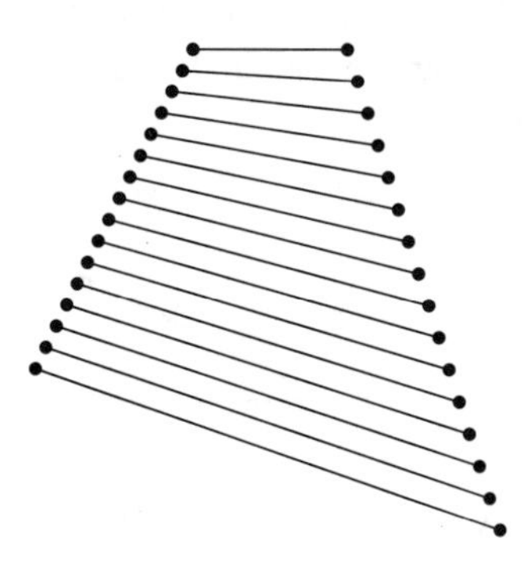

1.6 Area

50. A Japanese Temple Problem

"During the greater part of the Edo period (1603–1867) Japan was almost completely cut off from the western world. Books on mathematics, if they entered Japan at all, must have been scarce, and yet, during this long period of isolation people of all social classes, from farmers to samurai, produced theorems in Euclidean geometry which are remarkably different from those produced in the west during the centuries of schism, and sometimes anticipated these theorems by many years.

"These theorems were not published in books, but appeared as beautifully coloured drawings on wooden tablets which were hung under the roof in the precincts of a shrine or temple."

—H. Fukagawa and D. Pedoe,
preface to Japanese Temple Geometry Problems [FP]

Five squares are arranged as in the diagram. Show that the area of square S is equal to the area of triangle T.

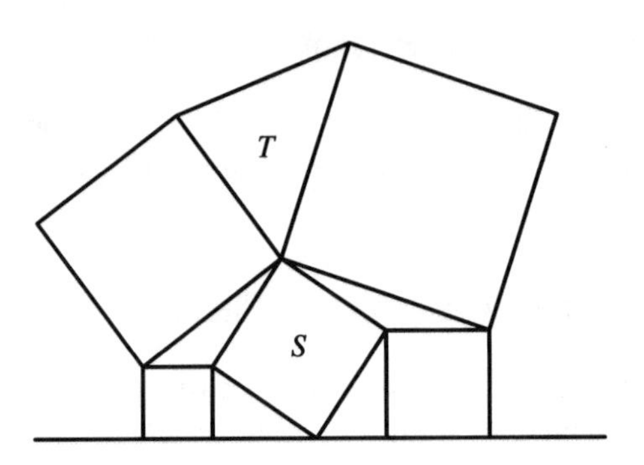

51. One Quadrilateral Begets Another

Suppose $ABCD$ is a convex quadrilateral and OA', OB', OC', and OD' are segments that are parallel to and equal in length to AB, BC, CD, and DA, respectively, as in the diagram. Determine the ratio of the area of $A'B'C'D'$ to the area of $ABCD$.

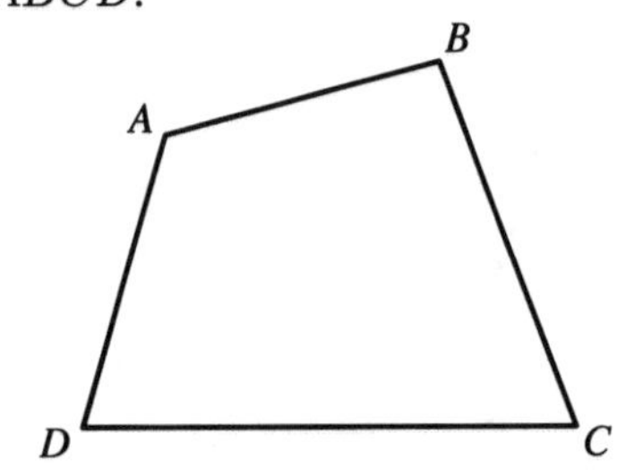

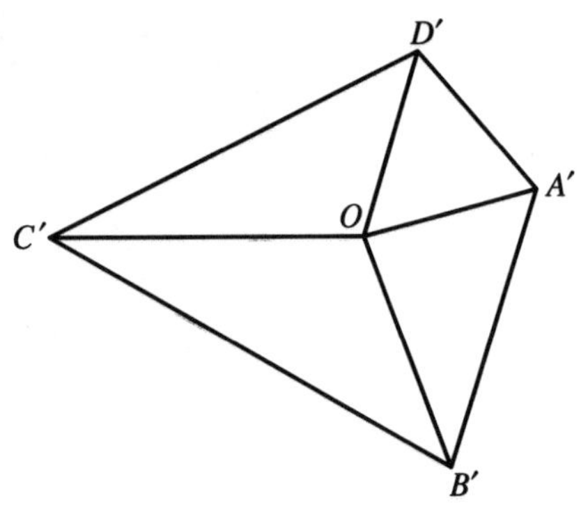

52. Rectangles in a Rectangle

In the figure, $ABCD$, $EFGH$, and $IFJH$ are all rectangles. Show that the sum of the areas of $EFGH$ and $IFJH$ is equal to the area of $ABCD$.

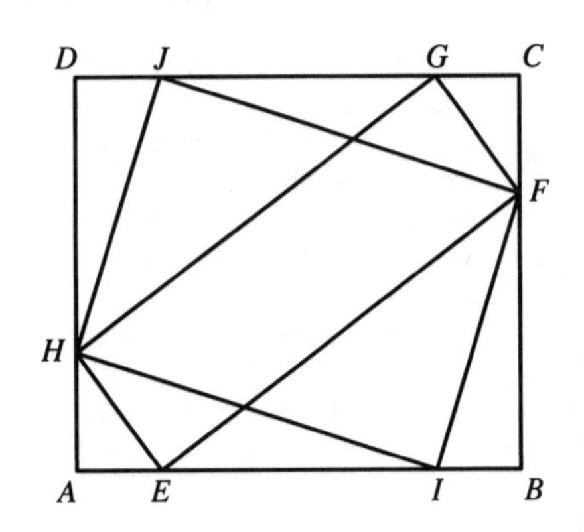

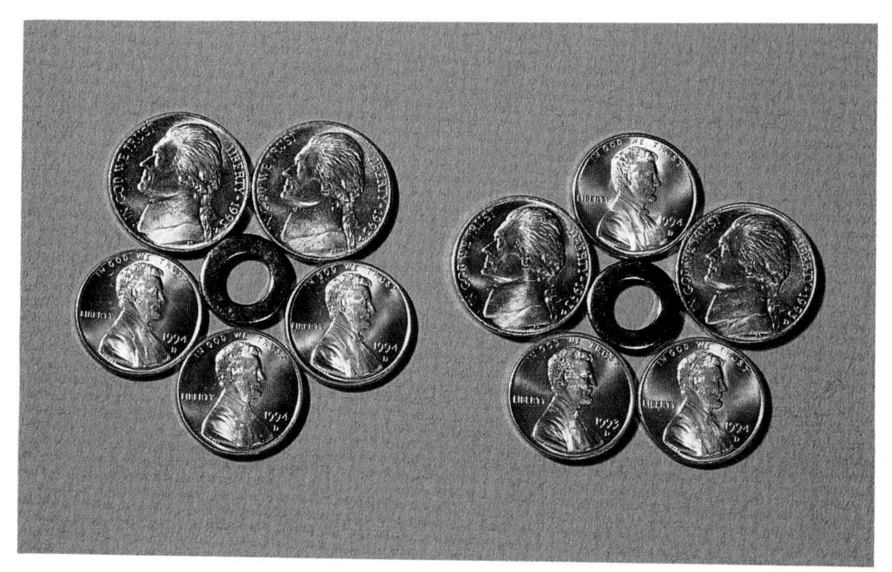

PLATE 1

There are two ways of arranging three pennies and two nickels around a central disk: nickels together (left) or nickels apart (right). In one case the central disk is larger than the other, though the difference is amazingly small. Problem 43 asks you to determine which case has the larger central disk.

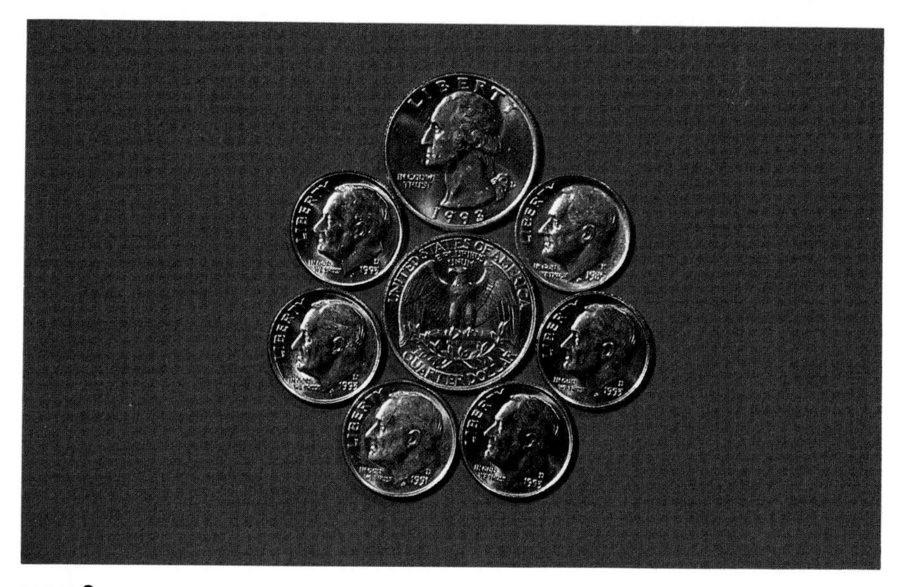

PLATE 2

If minting tolerances are taken into account, then six dimes and a quarter fit exactly around another quarter, as discovered by Garry Ford.

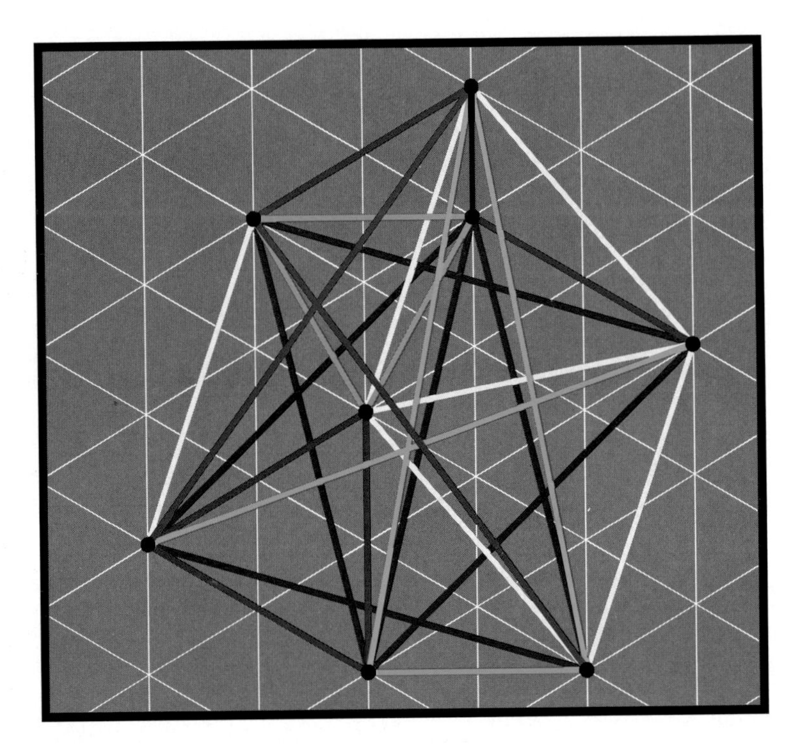

PLATE 3

This configuration of eight points, discovered by Ilona Palásti in 1989, is the largest known with the property that distinct distances occur once, twice, three times, and so on; see Problem 68. In this example, one distance occurs once (black), one distance occurs twice (red), one occurs three times (cyan), one occurs four times (green), one occurs five times (magenta), one occurs six times (yellow), and one occurs seven times (blue).

53. A Constant Difference

Suppose P, Q, and R are arbitrary points on the sides BC, CD, and DA, respectively, of the parallelogram $ABCD$. Join A to P to R to B to Q to A to form a star polygon $APRBQ$. The region outside the star polygon and inside the parallelogram is colored red. The pentagonal region inside the star polygon and bounded by the lines AP, BQ, PR, QA, and RB is colored blue. Show that the difference in areas between the red and blue region is independent of the choice of points P, Q, and R.

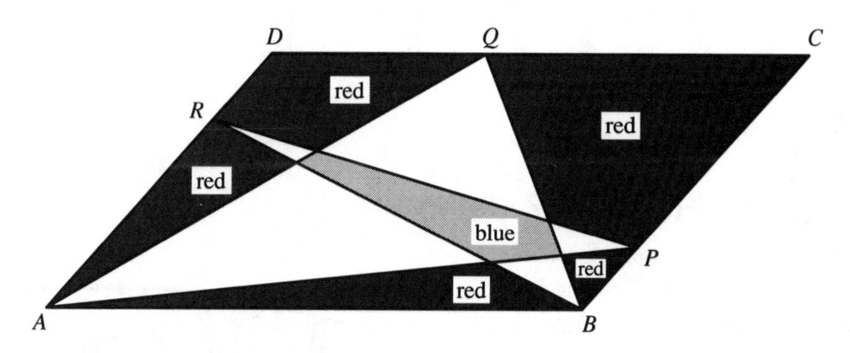

54. An Area Comparison

Let P be any point inside an equilateral triangle ABC, and let D, E, and F be the reflections of P in the sides BC, AC, and AB respectively. Which is larger, the area of $\triangle ABC$ or the area of $\triangle DEF$?

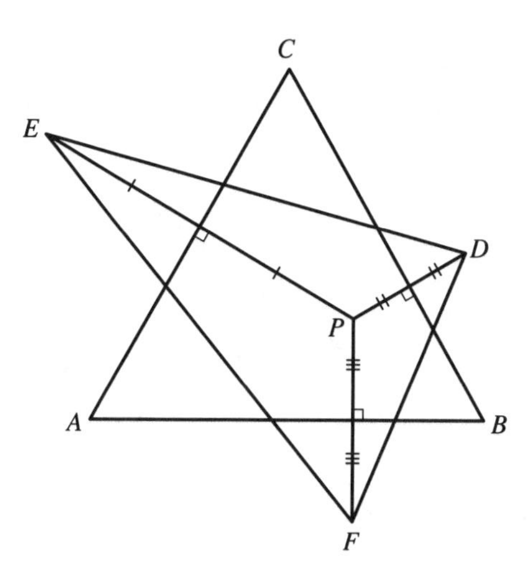

55. Another Area Comparison

In the figure, ABC is a right triangle, and $BCDE$ and $ACFG$ are squares. Which has a larger area, triangle ABH or quadrilateral $HICJ$?

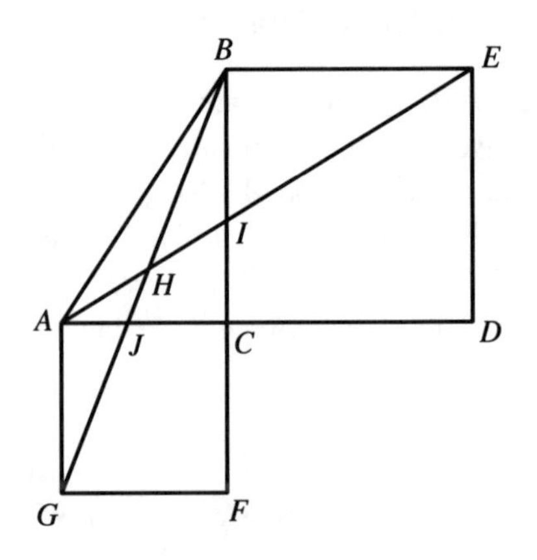

56. A Quadrilateral Inside a Parallelogram

In the diagram, the area of quadrilateral $WXYZ$ is equal to one-half that of the parallelogram $ABCD$. Prove that at least one of the diagonals of the quadrilateral is parallel to a pair of sides of the parallelogram.

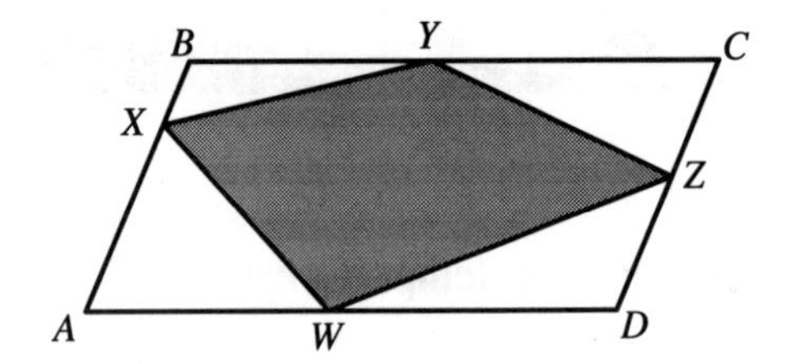

57. A Triangle in a Triangle

The area of $\triangle ABC$ is 1. Point D on BC is one third of the way from B to C, E is one third of the way from C to A, and F is one third of the way from A to B. Find the area of $\triangle GHI$.

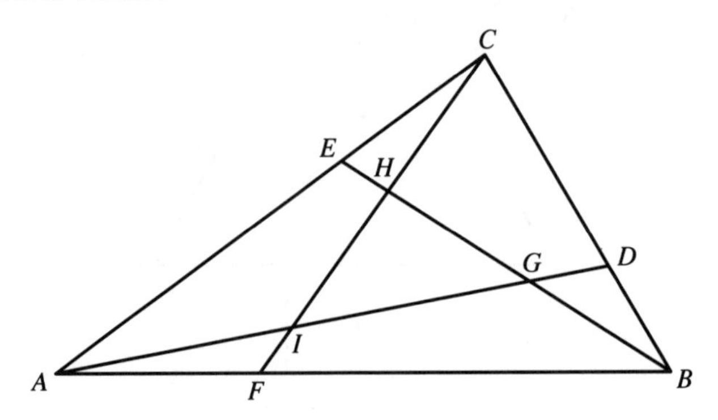

58. A Hexagon of Centroids

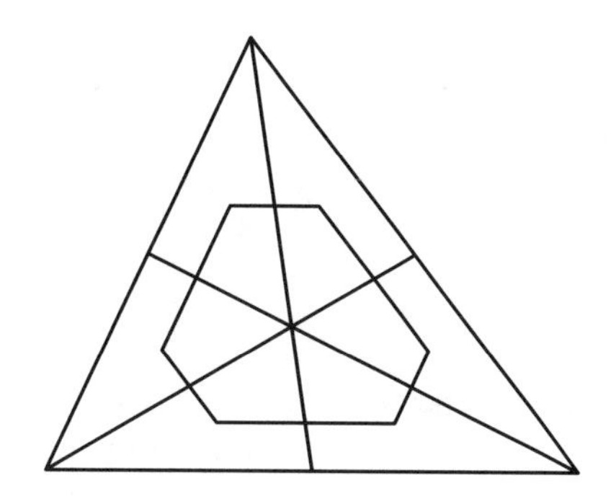

Line segments drawn from the vertices of a triangle to the midpoints of the opposite sides intersect at the centroid of the triangle and cut the triangle into six smaller triangles. If the area of the original triangle is 1, find the area of the hexagon whose vertices are the centroids of the six smaller triangles.

59. Five Triangles in a Pentagon

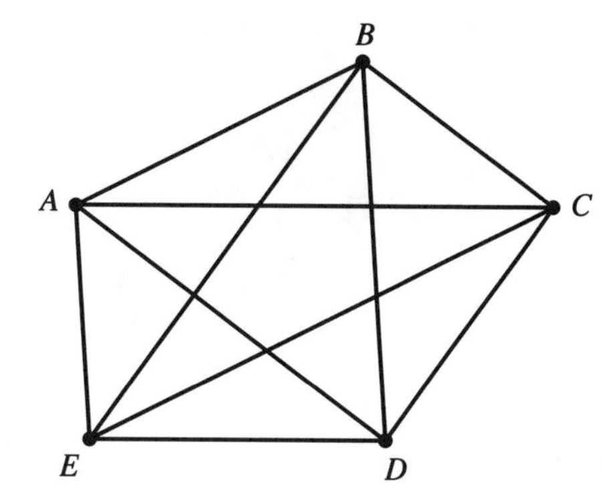

Suppose $ABCDE$ is a convex pentagon (not necessarily regular) with the property that each of the five triangles ABC, BCD, CDE, DEA, and EAB has area 1. What is the area of the pentagon?

60. Malfatti's Mistake

In the diagram, three circles are inscribed in an equilateral triangle in two different ways. In which case is the combined circular area greater?

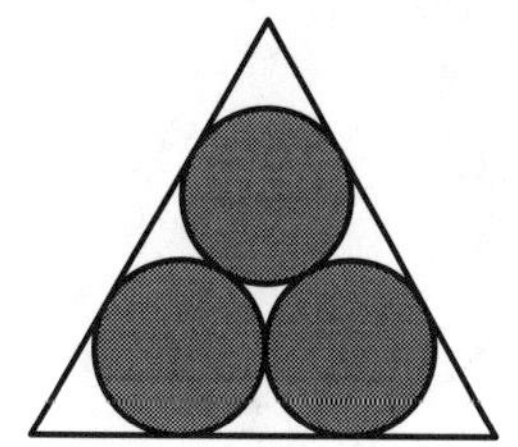

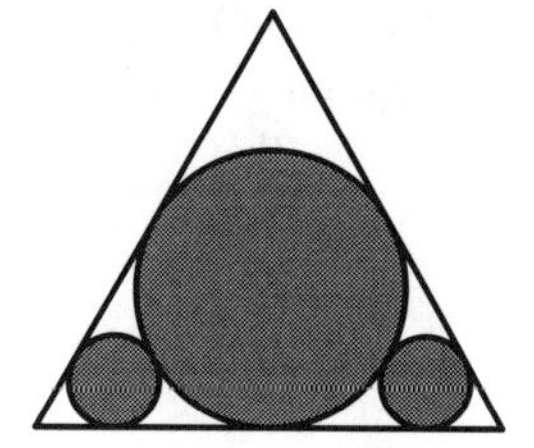

61. A Short Bisector

Find the shortest line segment that cuts a 3-4-5 right triangle into two pieces of equal area.

62. The Pizza Problem

Draw three chords through an arbitrary point inside a circle so that they make six 60° angles at the point and color the resulting "pizza slices" alternately black and white. Is the black area necessarily equal to the white area? What if four chords are used, making eight 45° angles?

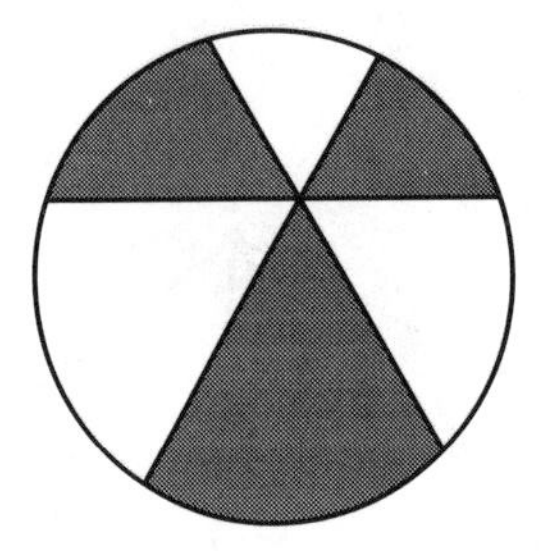

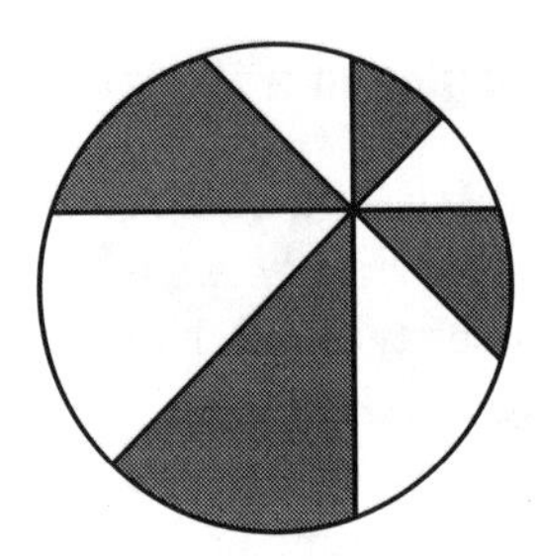

63. More Pizza!

> Sometimes miscommunication leads to new results. When one of the authors communicated the pizza problem (Problem 62) to noted problemist Murray Klamkin, he interpreted it slightly differently and came up with the following variation.

Suppose n is even and $n > 2$. Choose n equally-spaced points along a circle, connect each of them to an arbitrary point P inside the circle, and color the resulting regions alternately black and white (the diagram illustrates the $n = 8$ case). Prove that the black region equals the white region.

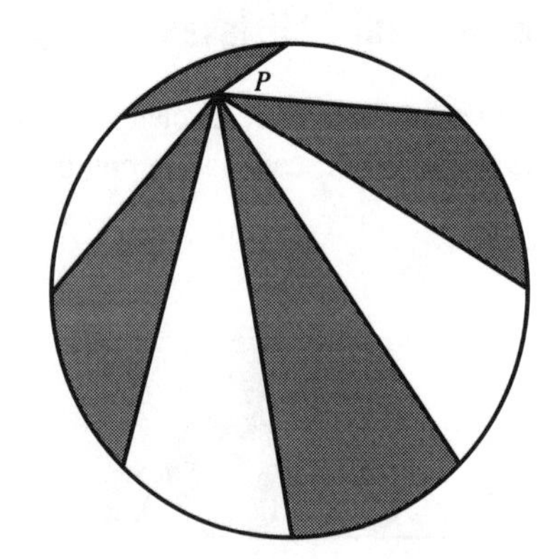

1.7 Miscellaneous

64. Nearest Neighbors

A collection of points in a plane is such that no two distances between pairs of points are equal. Show that if each point is connected to the point nearest it with a line segment, then no point is connected to more than five others.

65. Two-Distance Sets

Let us say that a set of points in the plane is a k-distance set if the distances between pairs of points in the set take on exactly k different values. For example, any set of two points is a 1-distance set, as is any set consisting of the vertices of an equilateral triangle. In fact, it is not hard to see that these are the only 1-distance sets. The vertices of an isosceles triangle form a 2-distance set. How many other 2-distance sets can you find?

66. An Almost-Symmetric Hexagon

$ABCDEF$ is a convex hexagon with each interior angle equal to 120°. Show that

$$AB - DE = CD - FA = EF - BC.$$

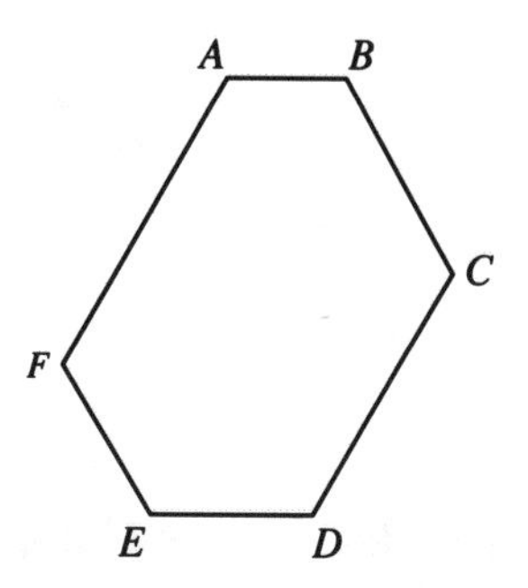

67. An Eight-Point Arrangement

Find an arrangement of eight points in the plane so that the perpendicular bisector of the line segment joining any two of the points passes through exactly two of the points.

68. Distances Determined by Five Points

In the diagram, four points, not all on one circle and no three on one line, are arranged so that they determine three distances, one three times, one twice, and a third just once. Can you arrange five points, no three on a line and no four on

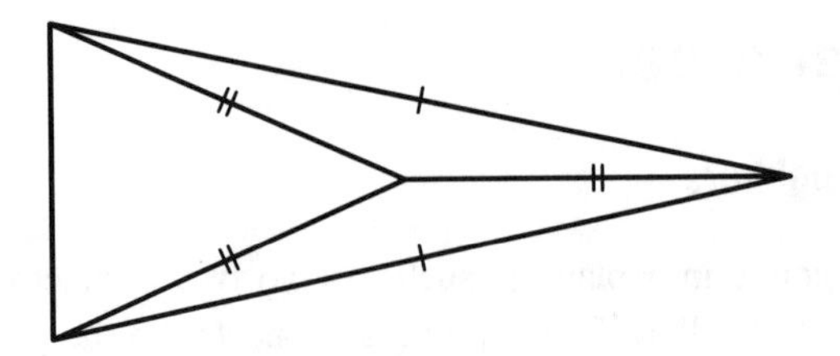

one circle, so that they determine one distance four times, a second distance three times, a third twice, and a fourth just once?

69. What's the Angle?

In the figure, $AB = BC = CD = DE = EF = FG = GA$. Find the measure of \ DAE.

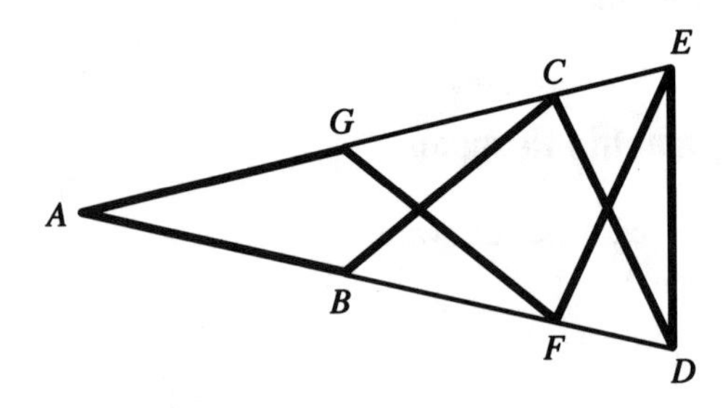

70. Find the Common Point

Given an angle at O and a constant k, let A and B be on the two rays forming the angle so that $1/OA + 1/OB = k$. Show that all the lines AB have a point in common.

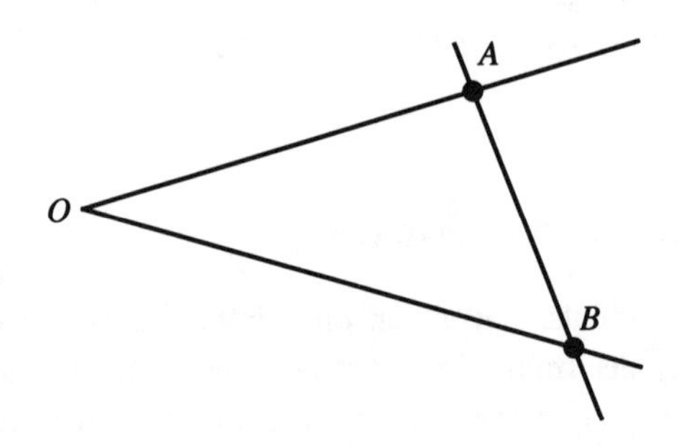

71. A Shady Garden Wall

Imagine that the point P in the sketch is a light source, and the polygon $ABCD$ is a closed garden wall.

Clearly the wall CD is in shadow, as is part of the wall AD. Is it possible to make a polygonal garden wall and position a light source inside it so that part or all of every wall is in shadow?

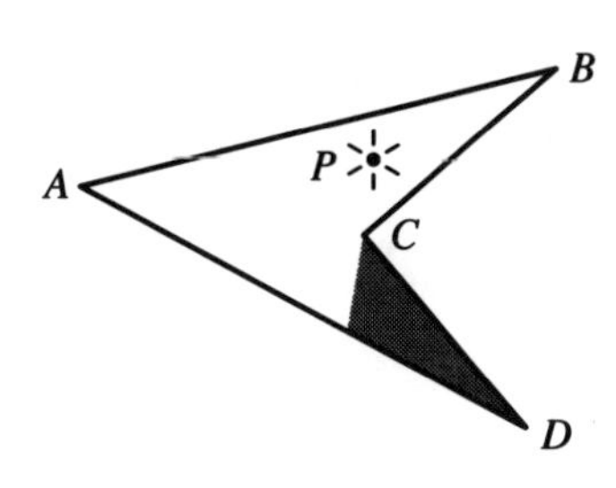

72. A Square Peg in a Nonround Hole?

A hole is cut in a piece of wood so that a square peg just fits through the hole in every orientation. In other words, no matter how the peg is oriented, all four corners of the peg touch the edge of the hole as the peg is pushed through the hole. Must the hole be a circle?

73. A Most Elementary Fact

In the sketch, $AB + BF = AD + DF$. Show that $AC + CF = AE + EF$.

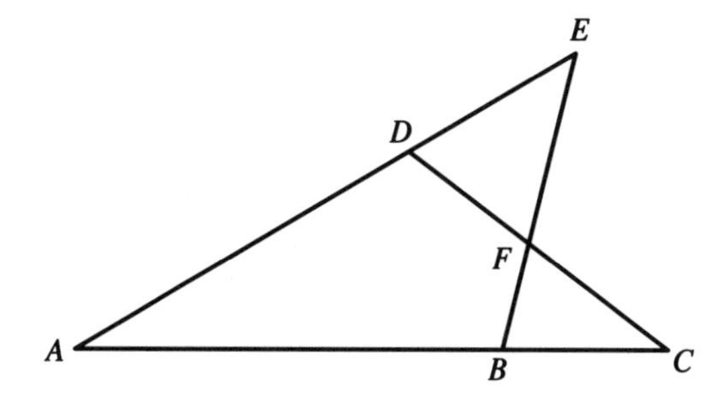

CHAPTER 2

Number Theory

2.1 Prime Numbers

74. No Primes Here

Consider the following infinite array of positive integers.

$$\begin{array}{cccccccc} 4 & 7 & 10 & 13 & 16 & 19 & \cdots \\ 7 & 12 & 17 & 22 & 27 & 32 & \cdots \\ 10 & 17 & 24 & 31 & 38 & 45 & \cdots \\ 13 & 22 & 31 & 40 & 49 & 58 & \cdots \\ \vdots & & & & & & \end{array}$$

The array continues in such a way that each row forms an arithmetic progression, and the ith column is the same as the ith row. Show that for every positive integer n, $2n+1$ is prime if and only if n does not appear in the array.

75. Squares, Primes, and Squares Plus Primes

Show that

- Infinitely many perfect squares are a sum of a perfect square and a prime number.
- Infinitely many perfect squares are not a sum of a perfect square and a prime number.

76. Not Goldbach's Conjecture

The infamous conjecture of Goldbach is the assertion that every even number greater than 2 is the sum of two primes. Except for 2, 4, and 6, every even integer is a sum of two positive composite integers: $n = 4 + (n - 4)$. What is the largest positive even integer that is not a sum of two odd composite integers?

77. They're There, but Where?

In 1772 Euler discovered the curious fact that $n^2 + n + 41$ is prime when n is any of $0, 1, 2, 3, \ldots, 39$. Show that there exist 40 consecutive integer values of n for which this polynomial is *not* prime.

78. A Pascal Pattern

It is remarkable that some simple facts about common objects can remain undiscovered for centuries. The following property of binomial coefficients was apparently not known until 1977, when it was discovered by Paul Erdős and George Szekeres.

Show that, excluding 1s, any two entries in the same row of Pascal's triangle have a common factor.

2.2 Digits

79. 1011 and All That

For small integers N it is easy to find integers whose base-10 representation contains only zeros and ones that are divisible by N. For example: 2 divides 110, 3 divides 111, 4 divides 100.

True or False: For each positive integer N there is at least one positive integer that consists wholly of zeros and ones and is divisible by N?

80. Only Ones

Note that 111 is divisible by 3. Show that there are infinitely many integers n such that the base-10 number consisting of n 1s is divisible by n.

81. Getting Ten Digits

True or False: Every positive integer has a multiple whose base-10 representation includes all ten digits 0, 1, 2, 3 4, 5, 6, 7, 8, 9?

82. Reversing Multiples

If we multiply the four-digit number 1089 by 9 we get 9801, a four-digit number containing the same digits as the original number—only in reverse! We might say that 1089 is *reversed* by multiplication by 9. Is there a four-digit number that is reversed by multiplication by 4?

83. Can You Beat a Billion?

Arrange the digits 0 through 9, using each exactly once, to form five two-digit numbers whose product is largest. For example, $10 \cdot 23 \cdot 45 \cdot 67 \cdot 89$ yields 61,717,050.

84. A Classy Social Security Number

Use the digits 1 through 9 once only to form a nine-digit number such that the first (leftmost) eight digits form a number divisible by 8, the first seven form a number divisible by 7, and so on. How many such numbers are there?

85. Three Repeated Digits in Two Bases

The number 40 has the unusual property that its base-7 representation is 55, while its base-9 representation is 44. Find the smallest positive integer that, in two bases, has representations consisting of three identical digits.

86. A Highly Divisible Number

What is the smallest positive integer that consists of the ten digits 0 through 9, each used just once, and is divisible by each of the digits 2 through 9?

87. Bilingual Palindromes

Some years, such as 1991, are palindromic: they read the same forwards or backwards. During 1991, the Hebrew calendar, which celebrates its New Year near the end of summer, was in the years 5751 and 5752, neither of which is a palindrome. Have there been any times since the year 0 that were palindromic in both the standard and the Hebrew calendars?

88. A Very Good Year

The year 1979 was unusual in that it results from stringing together distinct two-digit primes, namely 19, 97, and 79. The next time this happens is in 2311. When will this happen for the last time? Remember, the primes must be distinct, and there can be more than three of them.

89. Digital Diversity and Unbiased Numbers

A positive integer is said to be *digitally diverse* (DD) if the digits in its decimal representation are all different; for example, 415 is DD but 414 is not. A positive integer is said to be *unbiased* if exactly half the positive integers less than it are DD. Determine all the unbiased numbers.

90. A Really Big Number

A million is a big number. And the factorial operation makes even small numbers big. Consider 1,000,000!, a truly gigantic number. What is its rightmost nonzero digit?

2.3 Diophantine Equations

91. The Careless Dealer

While dealing, South dropped several of the undealt cards. He observed that the number of cards on the floor was two-thirds of the number he had already dealt to West, and that the number already dealt to East was two-thirds of the number of undealt cards he still held in his hand. How many cards had been dealt? (Note that the cards are dealt in the order West, North, East, South.)

92. Three Numbers and their Cubes

Find all integer solutions to the system of equations:

$$x^3 + y^3 + z^3 = 3 = x + y + z.$$

93. A Sum of Fractions

(a) Show that the equation $x/y + y/z + z/x = 1$ has no solutions in positive integers.

(b) Does the equation $x/y+y/z+z/x = 2$ have any solutions in positive integers?

94. Quadruplets with Square Triplets

Find four distinct positive integers such that the sum of any three of them is a perfect square.

95. When Does The Perimeter Equal The Area?

The right triangle with side-lengths 5, 12, and 13 has perimeter 30 and area 30. There is only one other right triangle with integer side-lengths having equal numerical values for area and perimeter. Find it.

96. Amicable Rectangles

If R and S are two rectangles with integer sides such that the perimeter of R equals the area of S and the perimeter of S equals the area of R, call R and S an *amicable* pair of rectangles. For example, a 4×6 and a 2×10 rectangle are amicable. Find all amicable pairs of rectangles.

97. Strange Boxes

A box with dimensions $5 \times 5 \times 10$ has the strange property that its surface area, 250, is numerically equal to its volume. Find all boxes with integer sides that have this property.

98. Dividing a Product by a Sum

Find all positive integers N for which the product $1 \cdot 2 \cdot 3 \cdots N$ is divisible by the sum $1 + 2 + 3 + \cdots + N$.

99. They're in the Money

Alice and Bob entered a bank and cashed checks for the same amount. They received the proceeds in pennies, half-dollars, and silver dollars, but not in the same combination. However, they each received the same total number of coins. What is the smallest possible value for the amount of each check?

100. Lucky Numbers

Call an integer *lucky* if it is a sum of positive integers (not necessarily distinct) whose reciprocals sum to 1. For example, 4 and 11 are lucky: $4 = 2 + 2$ and $\frac{1}{2} + \frac{1}{2} = 1$, and $11 = 2 + 3 + 6$ and $\frac{1}{2} + \frac{1}{3} + \frac{1}{6} = 1$. But 2, 3, and 5 are unlucky. How many unlucky numbers exist?

101. Abnormal Deviations

The mean and standard deviation of any set of seven consecutive integers are both integers. For example, the mean of 6, 7, 8, 9, 10, 11, and 12 is 9, and the standard deviation is 2. Find the next positive integer k such that the mean and standard deviation of any set of k consecutive integers are both integers.

Note. The standard deviation of a set of numbers $x_1, x_2, \dots, x_k$ is given by the formula

$$\sigma = \sqrt{\frac{\sum_{i=1}^{k}(x_i - \overline{x})^2}{k}},$$

where $\overline{x}$ is the mean of the numbers.

102. Problem for a New Millenium

For how many positive integers n is $\left(1999 + \frac{1}{2}\right)^n + \left(2000 + \frac{1}{2}\right)^n$ an integer?

103. Using Unit Fractions to Approximate a Unit

For which four distinct positive integers m, n, p, q is the sum $\frac{1}{m} + \frac{1}{n} + \frac{1}{p} + \frac{1}{q}$ less than 1 but as close as possible to 1?

104. A Necessary Square?

True or False: If a, b, and c are positive integers such that no integer greater than 1 divides all of them and if $\frac{1}{a}+\frac{1}{b}=\frac{1}{c}$, then $a+b$ is a perfect square? For example, $\frac{1}{3}+\frac{1}{6}=\frac{1}{2}$ and $3+6=3^2$.

2.4 Sets of Numbers

105. Fibonacci Magic

Is there a 3×3 magic square consisting of distinct Fibonacci numbers (both f_1 and f_2 may be used; that is, two 1s are allowed)?

Note. The Fibonacci numbers f_n are defined by $f_0 = 0$, $f_1 = 1$, and $f_{n+2} = f_n + f_{n+1}$. The first few are $0, 1, 1, 2, 3, 5, 8, 13, \ldots$. A magic square has the property that the eight sums along rows, columns, and the two diagonals are all the same number.

106. A Property of {2, 3, 5}

True or False: The set $\{2, 3, 5\}$ is the only triple of distinct positive integers such that the product of any two members leaves a remainder of one when divided by the third?

107. An Odd Set of Positive Integers

Is there a set S of positive integers such that a number is in S if and only if it is a sum of two distinct members of S or a sum of two distinct positive integers not in S?

108. Sums of Squares

Some integers can be written as a sum of two or more distinct positive squares and some cannot. For example, $42 = 1 + 16 + 25$ and $73 = 9 + 64$, but neither 43 nor 72 can be written as a sum of distinct positive squares. There is a largest integer that cannot be written as a sum of distinct positive squares. Find it.

109. Sums and Differences

Let $A = \{a_1, a_2, \ldots, a_n\}$ be a set of distinct positive integers. Let S be the number of distinct integers of the form $a_i + a_j, i, j = 1, 2, \ldots, n$; let D be the number of distinct integers of the form $a_i - a_j, i, j = 1, 2, \ldots, n$. For example, if $A = \{1, 2, 4\}$ then $S = 6$ (sums are 2, 3, 4, 5, 6, 8) and $D = 7$ (differences are -3, -2, -1, 0, 1, 2, 3). Prove or disprove: D is never less than S.

110. A Radical Equation

Solve the equation

$$\frac{1 - \sqrt{2} + \sqrt{3}}{1 + \sqrt{2} - \sqrt{3}} = \frac{\sqrt{x} + \sqrt{y}}{2},$$

where x and y denote nonnegative integers.

CHAPTER 3

Algebra

3.1 Polynomials

111. An Elusive Quadratic

Find a quadratic polynomial with integer coefficients $p(x) = ax^2 + bx + c$ such that $p(1)$, $p(2)$, $p(3)$, and $p(4)$ are perfect squares, but $p(5)$ is not.

112. A Polynomial Fitting Problem

The polynomial $p(x) = \frac{1}{2}x^2 - \frac{1}{2}x + 2$ has the property that $p(1) = 2$, $p(2) = 3$, and $p(3) = 5$. Can you find a polynomial $q(x)$ with *integer* coefficients such that $q(1) = 2$, $q(2) = 3$, and $q(3) = 5$?

113. A Polynomial Oddity

Suppose $P(x)$ is a polynomial with integer coefficients and $P(a) = P(b) = P(c) = P(d) = 2$ for distinct integers a, b, c, and d. Which of the following is true?

(a) There is no integer k such that $P(k) = 1$, 3, 5, 7, or 9.

(b) There is an integer k such that $P(k) = 1$, 3, 5, 7, or 9.

(c) Neither (a) nor (b) is true in general; there are examples for both.

114. A Polynomial Pattern

Suppose $P(x)$ is a polynomial of degree 8 with real coefficients and $P(k) = \frac{1}{k}$ for $k = 1, 2, 3, \ldots, 9$. Determine the number $P(10)$.

115. Powerful Patterns

Suppose that a, b, and c satisfy the equations

$$a+b+c=3,$$

$$a^2+b^2+c^2=5,$$

$$a^3+b^3+c^3=7.$$

Find $a^4+b^4+c^4$.

3.2 Miscellaneous

116. Christmas Time

On Christmas Eve, as Santa Claus left his workshop with his sleigh full of toys, he noticed that the second hand of his watch was directly over the 12. After traveling 8 miles, he looked at his watch again and observed that the minute hand and the hour hand coincided. If Santa's average speed over the 8 miles was 33 miles per hour, at what time had he left the workshop?

117. Play Ball!

After hitting a single, a certain baseball player observed that his batting average went up by exactly 10 points. If this was not his first hit, how many hits does he now have?

Note. A batting average is the quotient: number of base hits/number of at-bats; it is a real number between 0 and 1 and is traditionally given to three significant digits. An increase of ten "points" refers to an increase in this average of exactly 10/1000.

118. The Potato Peelers

Alice and Bob are on kitchen duty, each peeling potatoes at the rate of one per minute. They start with the same number, but Alice throws one unpeeled potato onto Bob's pile after every second one she peels. At a certain moment, Bob has twice as many potatoes to be peeled as Alice. Five minutes later this ratio has increased to 7 : 3. When will the ratio be 3 : 1? (Assume that Alice's completion of the peeling of a potato and her tossing of a potato onto Bob's pile occur simultaneously at the end of each second minute.)

119. Playing the Stock Market

Alice and Bob work for the Acme Widget Company, and each participates in one of the company's stock purchase plans. Over a period of n months, Alice invests a fixed number of dollars each month toward the purchase of shares of Acme stock, while Bob purchases a fixed number of shares of Acme stock each month. Since the price of the stock fluctuates over time, Bob may invest a different number of dollars each month, and Alice may purchase a different number of shares each month, and may sometimes purchase fractional shares. Assuming no brokerage or other fees, for whom is the average cost per share smaller?

120. An Impossible Inequality

Prove that there do not exist real numbers x, y, and z such that

$$-2 < \frac{x^2 + y^2 + z^2}{xy + yz + xz} < 1.$$

121. Radical Multiples

The integer nearest to $\sqrt{2}$ is 1. If we square $1 + \sqrt{2}$ we get $3 + 2\sqrt{2}$, and 3 is the integer nearest to $2\sqrt{2}$. If we cube $1 + \sqrt{2}$ we get $7 + 5\sqrt{2}$, and 7 is the integer nearest to $5\sqrt{2}$. Show that if $(1 + \sqrt{2})^n = P + Q\sqrt{2}$, where n, P, and Q are positive integers, then P is the integer nearest to $Q\sqrt{2}$.

122. Pascal's Determinant

Suppose that Pascal's triangle is rearranged as follows:

$$\begin{array}{cccccc} 1 & 1 & 1 & 1 & 1 & \cdots \\ 1 & 2 & 3 & 4 & 5 & \cdots \\ 1 & 3 & 6 & 10 & 15 & \cdots \\ 1 & 4 & 10 & 20 & 35 & \cdots \\ 1 & 5 & 15 & 35 & 70 & \cdots \\ \vdots & & & & & \end{array}$$

The first row and column in this array consist entirely of 1s, and every other number is the sum of the number to its left and the number above it. Show that for every positive integer n, the determinant of the matrix consisting of the first n rows and columns of this array is 1.

123. A Shuffling Reconstruction

A device that shuffles cards always rearranges them in the same way relative to the order in which they are placed into it. The ace through king of hearts are arranged in order with the ace on top and the king on the bottom. After two shuffles, the order of the cards, from top to bottom, is

10, 9, Q, 8, K, 3, 4, A, 5, J, 6, 2, 7.

What was the order of the cards after the first shuffle?

124. Associativity and Commutativity

Suppose $*$ is an associative operation on a set S, and A, B, and C are three elements of S such that

$$A * B = B * A,$$

$$A * C = C * A,$$

$$A * B * A = A, \quad \text{and}$$

$$B * A * B = B.$$

Prove that $B * C = C * B$.

Note. To say that $*$ is associative means that $X * (Y * Z) = (X * Y) * Z$ for all X, Y, and Z in S.

CHAPTER 4

Combinatorics and Graph Theory

4.1 Graphs and Networks

125. A Campus Stroll

On his daily campus stroll, Bob visited each building exactly twice using only the paths shown in the diagram. When he got back to his office, he told Alice that he had visited the buildings in the order

AONHKLECFBPIMGDJMLNHJKDGPBFIEOAC.

After a little thought, Alice said, "Bob, you're not quite right." Reconsidering, Bob responded, "I see that I've carelessly transposed two successive buildings at exactly one point in the order." Without knowing which building is which on the map, name the transposed buildings.

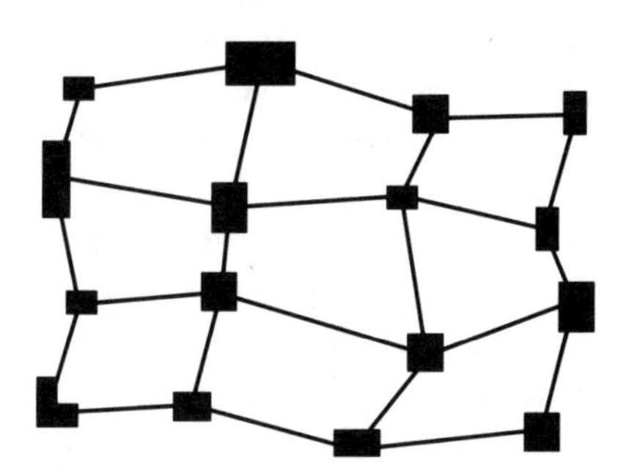

126. Visit these Places

Is there a path along the lines and arcs in the diagram that passes through each junction point once and only once?

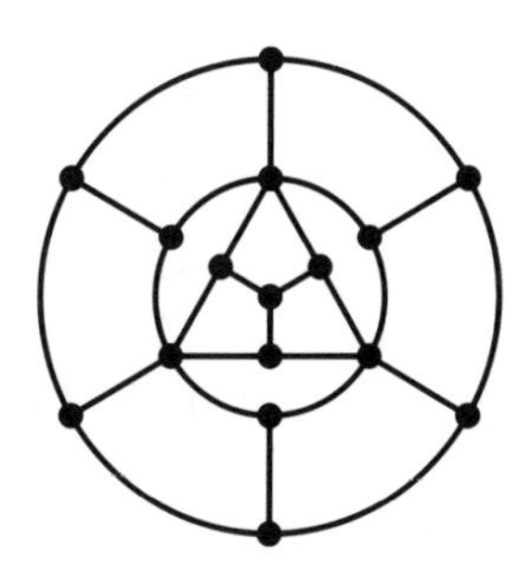

127. The Color of Money

A finite number of pennies are put on a table so that some of them touch, but none overlap. The pennies are then colored so that no two touching pennies are the same color. The famous four-color theorem implies that this can always be done with at most four colors. Is there an arrangement of pennies that requires four colors?

128. Playing with Matches

A matchstick graph is a set of unit-length sticks in the plane, with the proviso that two sticks may meet only at their endpoints. The points that are matchstick-ends are called the vertices of such a graph. For example, an equilateral triangle with unit side-length is a matchstick graph with 3 vertices in which each vertex is connected to two others. Find a matchstick graph in which each vertex is connected to exactly three other vertices. Try to minimize the number of vertices.

129. A Universal Coloring

Sixteen Easter eggs are arranged in a ring as illustrated.

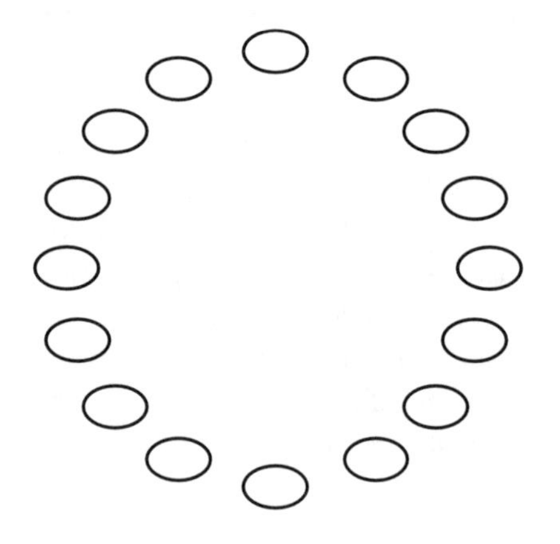

(a) Using four colors (red, yellow, green, blue), are you able to paint the eggs so that in going around the ring in either direction each of the sixteen possible color-pairs (red/red, blue/green, green/blue, and so on) occurs exactly once?

(b) What if there were only ten eggs and it was desired to get all ten unordered color pairs?

130. Hamiltonian Circuitry

Suppose that $2n$ points are evenly spaced around a circle and each pair of points is connected by a straight line segment, forming a complete graph with $2n$ vertices. A Hamiltonian circuit is a path that begins at some vertex in the graph and follows the edges of the graph, visiting all other vertices exactly once before returning to the starting vertex. Does this graph have a Hamiltonian circuit in which no two edges are parallel?

For example, the diagram shows a Hamiltonian circuit in the case $n = 3$. Notice that in this example the edges connecting 1 to 3 and 4 to 6 are parallel.

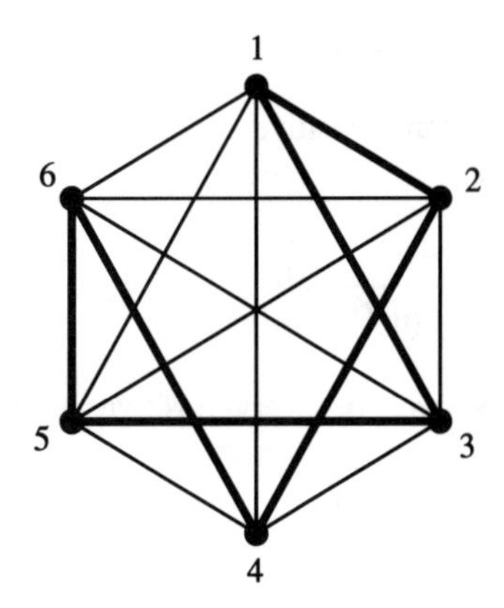

131. Cover the Honeycomb

(a) What is the maximum number of markers that can be placed on the hexagons in the figure so that no row (horizontal or slanted) of contiguous hexagons contains more than one marker?

(b) What is the minimum number of markers, no two in a row, that can be placed so that the addition of one more is impossible without having two markers in a row?

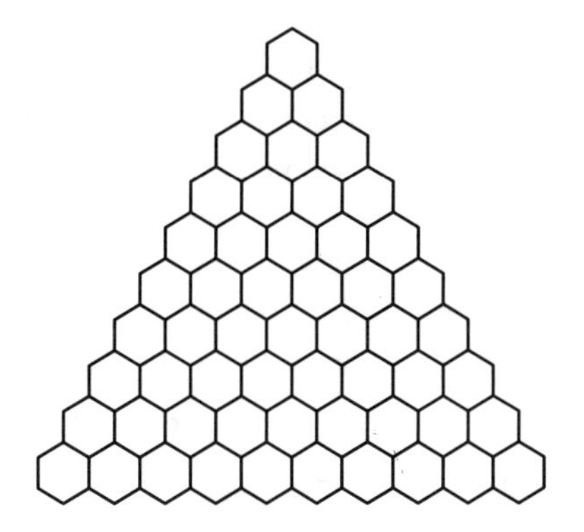

4.2 Counting

132. Can You Get There From Here?

Towns A, B, and C are connected to each other by several roads, with at least one road connecting each pair of towns. In going from A to B one can go directly on one of the roads joining them, or one can travel from A to C on one of the roads joining A and C and then travel from C to B on a road joining C and B. In all, there are 33 routes from A to B (including those via C). Similarly, there are 23 routes from B to C (including those via A). How many routes are there from A to C (including those via B)?

133. Decimal Diversity

How many positive integers have decimal representations consisting of distinct digits?

134. Digit Counting

Show that for every positive integer n, the total number of digits in the sequence $1, 2, 3, \ldots, 10^n$ is equal to the total number of zero digits in the sequence $1, 2, 3, \ldots, 10^{n+1}$.

135. How Long Would It Take You?

> Mathematically sophisticated customer: "I think a smart burglar could break this combination lock in a few minutes."
> Salesperson: "Oh, no! There are millions of combinations."

A commercially available combination door lock from the Simplex corporation has five buttons numbered 1 to 5. In a valid combination buttons may be pushed in groups, with each group being a simultaneous push of one or two or three or four or five buttons. Once pushed, a button stays down and cannot be pushed again. It is not necessary that all buttons be pushed. And the order within each group is irrelevant since the buttons in each group are pushed simultaneously. For example, the following are all valid combinations:

3 — 15 — 24 (same as 3 — 15 — 42) 3 — 1245
124 — 5 4 — 1 — 3 — 2 — 5 5
4 — 51 — 3 15 — 3 — 24 (different than first one above)

The combinations 12 — 24 and 14 — 23 — 1 are invalid because of the repetition. How many valid combinations are there?

136. How Many Triangles?

Given a regular polygon of n sides, how many triangles have vertices that are vertices of the polygon and have sides that are not sides of the polygon. For example, for a regular 7-gon there are seven such triangles.

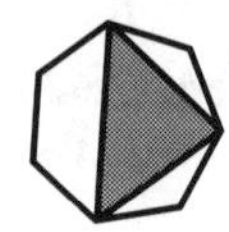 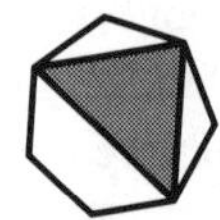 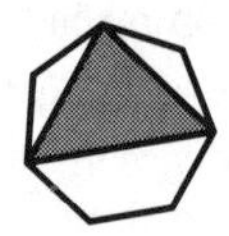    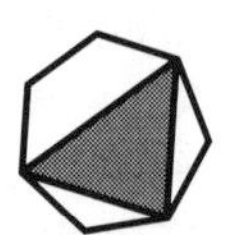

137. Breaking Up Space

A plane divides 3-space into two parts. Two intersecting planes separate 3-space into four parts. Three planes all passing through one point but not all passing through one line separate 3-space into eight parts.

(a) Into how many regions is 3-space divided by four planes all passing through one point, if no three of the planes pass through the same straight line?

(b) Same question as in (a), but with four replaced by an arbitrary positive integer.

138. A Piercing Diagonal

A diagonal line is drawn from one corner of an $m \times n$ grid of squares to the opposite corner. How many squares of the grid contain a segment of the line? For example, the figure shows that there are 12 squares in an 8×6 grid containing a segment of the diagonal.

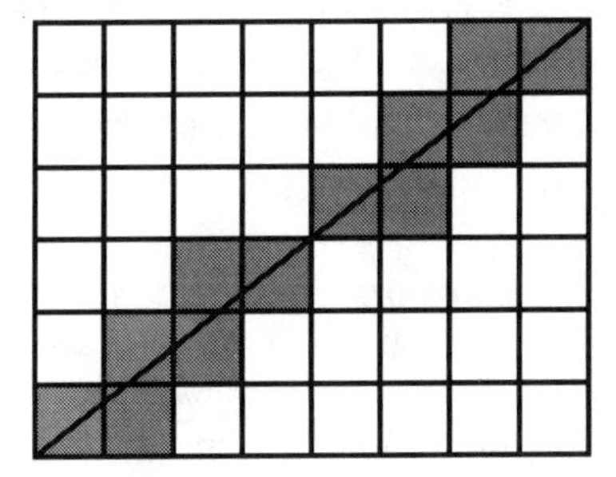

139. Can You Find the Key?

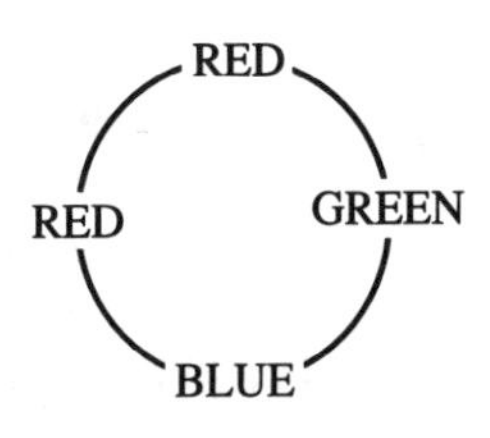

Suppose you have n keys on a circular key chain and wish to put a colored sleeve on each key so that it is identifiable by its color alone, without reference to its shape. For some n this can be done with fewer than n colors. For example, if you have 4 keys then you can place the colors on the keys as shown in the figure. The top key is "the RED key that is across from the BLUE key"; and the leftmost key is "the RED key that is adjacent to a BLUE one"; the other two are identifiable by their colors.

An identification scheme must work even if the key chain is flipped over, so one cannot use the words "right" and "left." Let $f(n)$ be the least number of colors necessary so that each key on a chain of n keys can be uniquely identified as explained above. Then $f(1) = 1$, $f(2) = 2$, and $f(3) = 3$. What is $f(139)$?

140. Count the Tilings

The diagram shows a tiling of a 2×7 rectangle with 1×1 and 1×2 tiles (singletons and dominoes; dominoes may be placed horizontally or vertically). How many such tilings of a 2×7 grid are there?

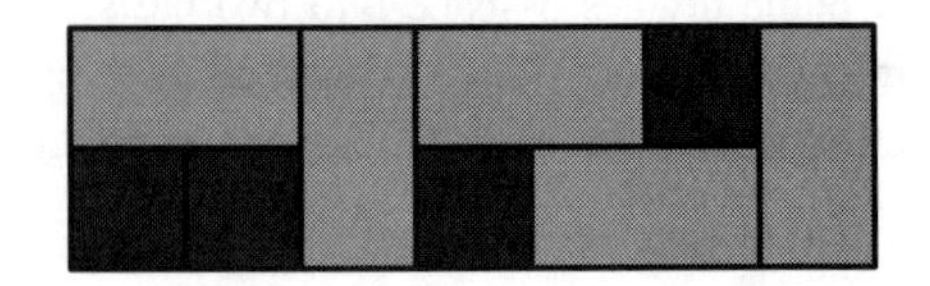

4.3 Miscellaneous

141. How to Win at the Lottery

A \$1 ticket in the Massachusetts Lottery consists of 6 different numbers chosen from $1, 2, 3, \ldots, 48$. On lottery day 6 different numbers from this set are chosen at random; a winning ticket is one that has at least 5 of these 6 numbers (order is irrelevant). Show that if one buys all tickets for which the sum of the entries is divisible by 47, then one is guaranteed of having a winner.

142. A Problem with the Elevators

In an apartment building there are seven elevators, each stopping at no more than six floors. If it is possible to go from any one floor to any other floor without changing elevators, what is the maximum number of floors in the building?

143. Avoiding Arithmetic Progressions

It is easy to split the integers $1, 2, 3, \ldots, 8$, into two sets such that neither contains a 3-term arithmetic progression: for example, $\{1, 2, 5, 6\}$ and $\{3, 4, 7, 8\}$ (there are two other ways). Suppose we can use three sets instead of two. How large a set $\{1, 2, \ldots, n\}$ can you split into three sets, none of which has a 3-term arithmetic progression?

144. Easy as One, Two, Three

Does any row of Pascal's triangle have three consecutive entries that are in the ratio $1 : 2 : 3$?

145. Difference Triangles

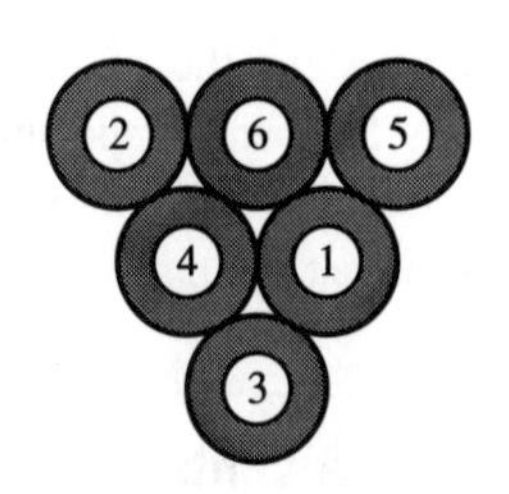

Colonel George Sicherman of Buffalo, while playing pool, wondered if the 15 balls could be arranged in a triangular array such that the number on each ball below the top row is the difference of the two numbers just above it. For 6 balls this is easy to do, as in the diagram.

Can such a difference triangle be formed using ten balls numbered 1 to 10 in a triangular array with 4 rows? What about Sicherman's original problem for 15 balls? And what about using $1, 2, 3, \ldots, 21$ to label balls in a triangle with six rows?

146. Ten Cannonballs

Ten congruent balls numbered 1 through 10 are to be stacked in pyramid fashion with six balls in the bottom layer, three in the middle layer, and one on top. After stacking, look at the set of differences obtained by subtracting labels of touching balls. For example, if the ball labeled 6 touches the one labeled 9, the difference is 3. Find a stacking for which the largest difference is as small as possible.

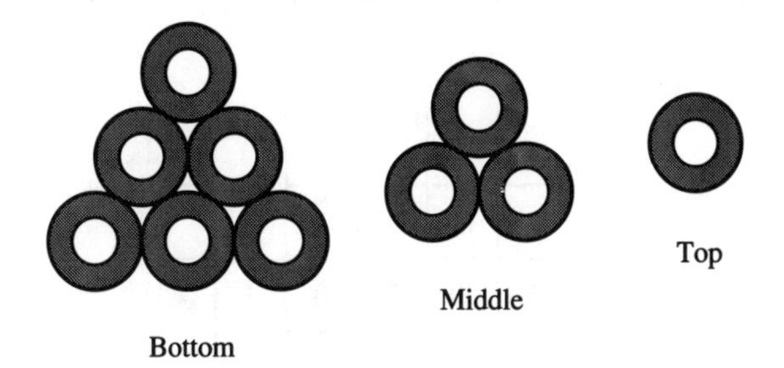

147. Christmas Confusion

Nine cards are arranged in a row, with 8 of them bearing words as in the diagram. Cards may be moved, one at a time, from a square to the then-vacant square, but

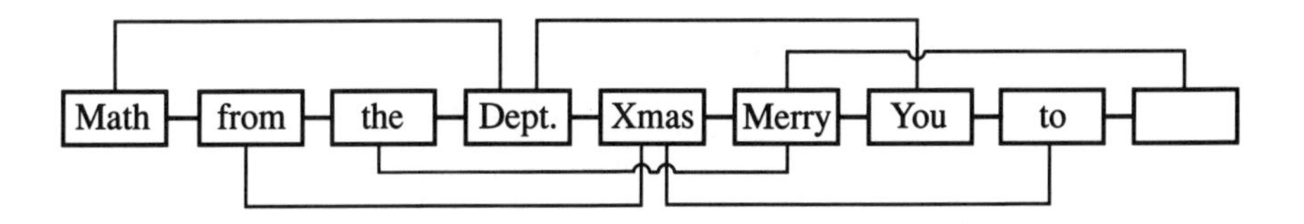

only along the 14 paths shown. If two cards may not occupy the same square at any time and if the space must end up at one of the ends, are you able to arrange the cards in the order: "Merry Xmas To You From the Math Department" with the space at the end? Try to use as few moves as you can.

148. A Veritable Babel

A college dormitory has 250 students. For every two students A and B, there is some language that A speaks that B doesn't speak, and some language that B speaks that A doesn't speak. What is the smallest total number of languages that could be known by the students?

149. Find the Pattern

The squares of an infinite chessboard are numbered as in the sketch. The number 0 is placed in the lower left-hand corner; each remaining square is numbered with the smallest nonnegative integer that does not already appear to the left of it in the same row or below it in the same column. If the first row (column) is called the zeroth row (column), which number will appear in the 378th row and 1707th column?

5	4	7	6	1	0	
4	5	6	7	0	1	
3	2	1	0	7	6	
2	3	0	1	6	7	
1	0	3	2	5	4	
0	1	2	3	4	5	

CHAPTER 5

Three-Dimensional Geometry

150. Only Isosceles Triangles

Arrange 8 points in 3-space so that for each of the 56 triples of points that they determine, at least two of the three distances between points in the triple are equal.

151. Finding the Middle Ground

The *midset* of the point sets S and T is the set of midpoints of all the line segments XY, where X is in S and Y is in T. If S and T are skew face diagonals of a cube (see diagram), what is the midset of S and T?

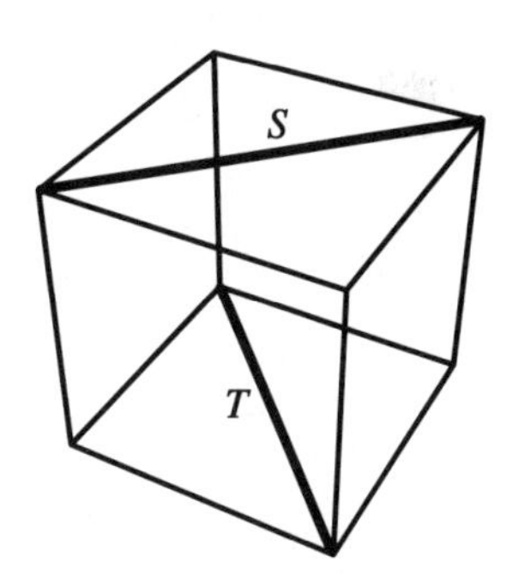

152. Spin the Rod

A rigid rod R, 14 feet long, is suspended horizontally by means of two vertical ropes 25 feet long, each of which is fastened to a ceiling and to an end of R. If R is twisted (that is, rotated about a vertical line through its center), it will rise. Through what angle must R be rotated (about a vertical line through its center) so that it rises exactly one foot?

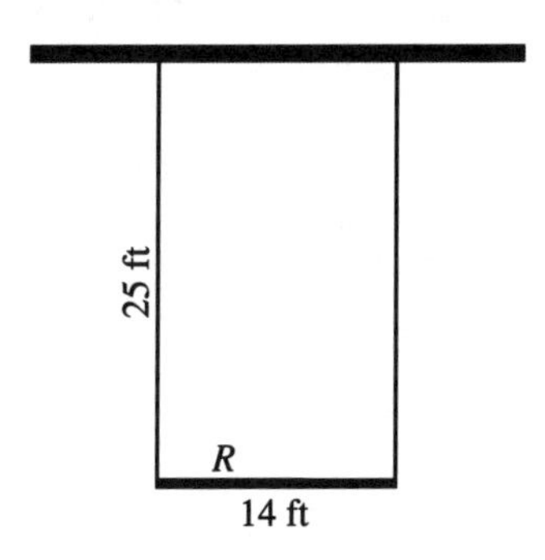

153. Traveling among the Pyramids

Find the length of the shortest route by which one can travel on foot from the top of one pyramid (A) to the top of the other (B). The pyramids have square bases and identical isosceles triangles for the other faces, and have the dimensions, in feet, as indicated in the diagrams.

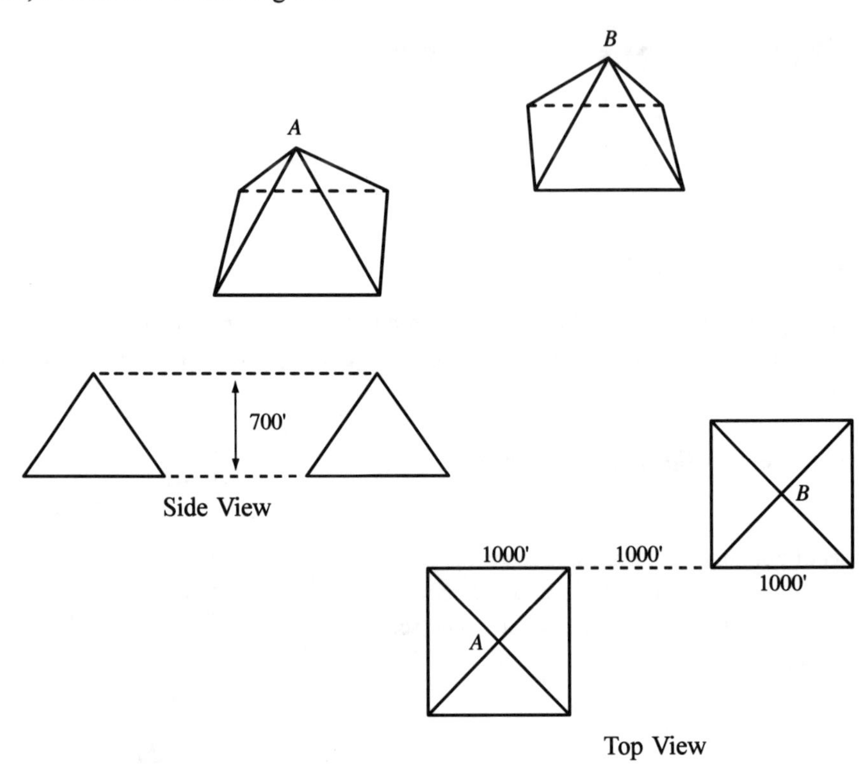

154. An Isosceles Tetrahedron

A tetrahedron is called *isosceles* if opposite edges have equal lengths, as in the figure. Prove that all the face angles of an isosceles tetrahedron must be acute.

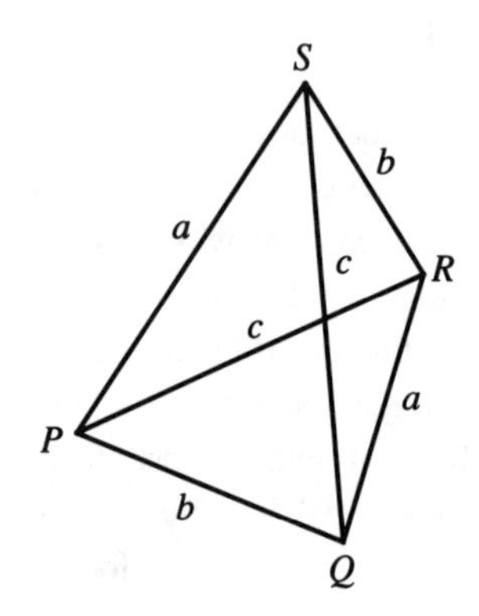

155. Connect the Dots

This beautiful problem, which has an intuitive appeal that can lead even professional geometers astray, is due to noted mathematician and expositor Victor Klee of the University of Washington.

For a set E in $\mathbb{R}^3$, let $L(E)$ consist of all points on all lines determined by any two points of E. Thus if V consists of the four vertices of a regular tetrahedron, then $L(V)$ consists of the six edges of the tetrahedron, extended infinitely in both directions, as illustrated.

True or False: Every point of $\mathbb{R}^3$ is in $L(L(V))$?

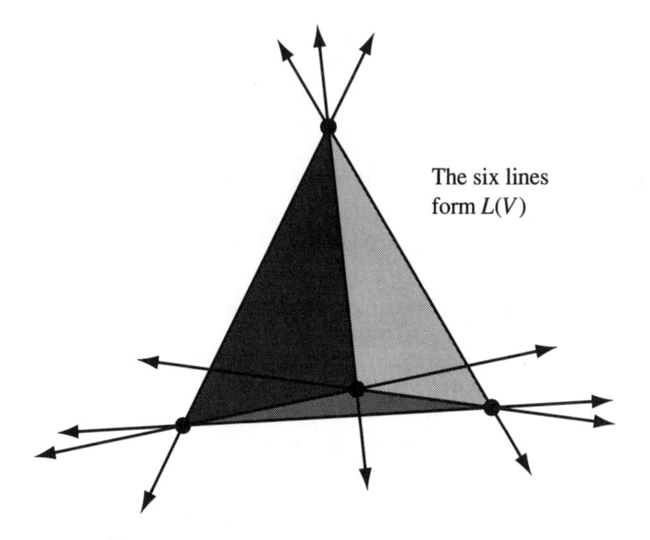

156. A Point and a Plane

From a fixed point not in the plane π, three mutually perpendicular line segments are drawn terminating in π. If a, b, c denote the lengths of the three line segments, show that the sum $1/a^2 + 1/b^2 + 1/c^2$ has the same value for all orientations of the three perpendicular lines.

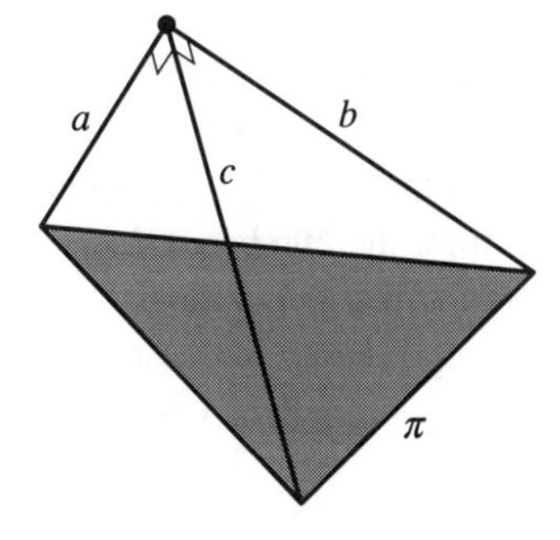

157. Painting a Cube

Some unit cubes are assembled to form a larger cube. Then some, but not all, of the faces of the large cube are painted. After the paint has dried, the large cube is disassembled and it is discovered that 218 of the unit cubes have some paint on them. What is the size of the large cube?

158. The Middle of a Moving Line

Suppose that L and M are two nonintersecting lines in 3-space with M being parallel to a line that is perpendicular to L. A line segment PQ of fixed length moves so that P is on L and Q is on M. What is the locus of the midpoint of PQ?

159. A Circle, a Sphere, and a Circle on a Sphere

A compass is used to draw a circle on a plane. Then, with no change in the opening, the compass is used to draw a circle on a sufficiently large sphere. Which is greater, the area of the circle in the plane or the area of that portion of the sphere enclosed by the circle on the sphere?

160. A New Dimension to a Famous Problem

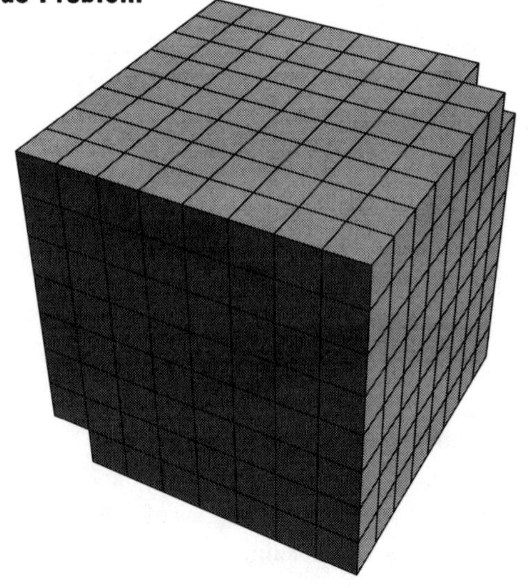

A famous problem asks whether an 8×8 chessboard with two opposite corners deleted can be tiled with dominoes, where a domino is a rectangle congruent to two adjacent squares of the board. The answer is no because each domino would have to cover one white and one black square, an impossibility since the number of white squares is different from the number of black ones. And using $1 \times n$ dominoes does not help because 62 is not divisible by any n in $\{3, 4, 5, 6, 7, 8\}$.

Now let C be an $8 \times 8 \times 8$ cube with two opposite corners removed. For which integers n is it possible to completely fill C using $1 \times 1 \times n$ boxes (in any orientation)?

161. Surveying on Earth

This problem, which is reminiscent of the classic "What color is the bear?" puzzle, seems impossible to answer at first but yet repays careful thought. Just ask yourself: "What color is their bear?" It is due to Lester R. Ford, an avid problemist who was editor of the *American Mathematical Monthly* during World War II and, later, president of the MAA.

Alice and Bob own roughly rectangular pieces of land on the planet Earth, which is assumed to be a perfect sphere of radius 3950 miles. Alice's land is bounded by four fences, two of which run in an exact north–south direction and two of which run in an exact east–west direction. Her north–south fences are exactly 10 miles long; her east–west fences are exactly 20 miles long. Bob's land is similarly bounded by four fences, but his north–south fences are 20 miles long and his east–west fences are 10 miles long. Whose plot of land has the greater area?

CHAPTER 6

Miscellaneous

162. A Puzzling Reflexicon

Lee Sallows (Nijmegen, The Netherlands), who has invented many beautiful self-referential objects, calls this pretty example a *reflexicon,* from *reflexive lexicon,* or self-determining word list.

Here is a self-enumerating crossword puzzle with a unique solution! Each of the six horizontal and six vertical entries is of the form, for example, "THIRTEEN NS." Here "THIRTEEN" can be any possible English number-word and "N" can be any letter in English. The idea is that if "THIRTEEN NS" is one of the entries, then the completed puzzle does indeed have *exactly* 13 instances of the letter "N." There are 12 entries, and so there will only be 12 different letters used in the completed puzzle. Every entry will have one blank cell, and an "S" occurs at the end when a plural necessitates it.

163. A Complicated Constant

Believe it or not, there are numbers a and b, with $a < b$, such that the expression

$$\sqrt{x+2\sqrt{x-1}}+\sqrt{x-2\sqrt{x-1}}$$

is constant for $a \leq x \leq b$. Find a and b, and the constant value of the expression.

164. Conservation of Blits

Imagine a 50×50 grid with the upper-left 25×25 quadrant containing some data (an integer, say) in each position and the rest blank. We wish to transform this grid into one that has the transpose of the 25×25 data array sitting in the lower-left quadrant using moves called "blits." A blit consists of selecting a rectangle of data and copying it by a single translation (no rotations or flips) to another location on the grid.

For example, one can transform

1	2		
3	4		

to

1	3		
2	4		

using 4 blits, since each blit could move just one entry at a time (a 1×1 rectangle). But one can do much better than 625 blits for the problem posed. How much better can you do?

165. The Electrician's Dilemma

A tunnel underneath a large mountain range serves as a conduit for 1001 identical wires; thus, at each end of the conduit, one sees 1001 wire-ends. Your job is to label all the ends with labels $\#1, \#2, \ldots, \#1001$, so that each wire has the same label at its two ends.

You may join together arbitrary groups of wires at either end; they will then conduct electricity through the join. Then you cross the mountains by a very expensive and dangerous helicopter ride to the other end, where you can feed electricity through any wire and check which of the other ends are live, attach notes to the wires, and make (or unmake) connections as desired. Then you fly back to the near end, perform the same sort of operations, fly back, and so on as often as required.

How can you accomplish your task with the smallest number of helicopter flights?

166. An Egg-Drop Experiment

Suppose we wish to know which windows in a 36-story building are safe to drop eggs from, and which will cause the eggs to break on landing. We make a few assumptions:

- An egg that survives a fall can be used again.
- A broken egg must be discarded.
- The effect of a fall is the same for all eggs.
- If an egg breaks when dropped, then it would break if dropped from a higher window.
- If an egg survives a fall, then it would survive a shorter fall.
- It is not ruled out that the first-floor windows break eggs, nor is it ruled out that the 36th-floor windows do not cause an egg to break.

If only one egg is available and we wish to be sure of obtaining the right result, the experiment can be carried out in only one way. Drop the egg from the first-floor window; if it survives, drop it from the second-floor window. Continue upward until it breaks. In the worst case, this method might require 36 droppings. Suppose two eggs are available. What is the least number of egg-droppings that is guaranteed to work in all cases?

167. The Race Goes to the Swiftest

Alice and Bob ran a marathon (assumed to be exactly 26.2 miles long) with Alice running at a perfectly uniform eight-minute-per-mile pace, and Bob running in fits and starts, but taking *exactly* 8 minutes and 1 second to complete each mile interval (this refers to all intervals of the form $(x, x+1)$, including, for example, the interval from 3.78 miles to 4.78 miles). Is it possible that Bob finished ahead of Alice?

168. A Competition Conundrum

In a recent athletic competition there were three participants, Alice, Bob, and Eve. In each event p_1 points were awarded for first place, p_2 points for second place, and p_3 points for third place, where $p_1 > p_2 > p_3 > 0$ and p_1, p_2, and p_3 are integers. There were no ties. Alice's total for all events was 22 points, while Bob and Eve each earned 9 points. Bob won the 100-yard dash. Who was second in the high jump? How many events were there in all?

169. A Messy Desk

Fifteen sheets of paper of various sizes and shapes lie on a desktop covering it completely. The sheets may overlap one another and may even hang over the edge of the desktop. Prove that five of the sheets can be removed so that the remaining ten sheets cover at least two-thirds of the desktop.

170. An Optimist and a Pessimist

Suppose $u_1, u_2, u_3, \ldots$ is a sequence of numbers in the open interval $(0, 1)$. Alice, an optimist who hopes to travel arbitrarily far, travels u_n inches on her nth step. Bob, a pessimist, is content to reach a chair one foot away and covers on his nth step a distance equal to u_n times the distance remaining between him and the chair. Show that if Alice can travel as far as she wishes, then Bob can get as close to the chair as he wishes.

171. The Ambiguous Clock

Imagine a clock for which the hour hand and minute hand have the same length. One can still tell time, most of the time. For example, at 6:00 one can be sure that the upper hand is the minute hand, for otherwise the upper hand ought to be halfway between two hours. How many times during a 12-hour period are the hands of such a clock in an ambiguous position as regards the time of day?

172. A Problemist's Joke

> The late Samuel Greitzer intensely disliked "new math" and would argue against including problems about liars and truthtellers, Venn diagrams, and non-base-10 arithmetic on math contests. As a joke, his friend Stanley Rabinowitz invented this problem, which makes use of all three concepts!

Alice and Bob live in the small principality of Binumeria, where everyone speaks only English and always tells the truth. But when it comes to numbers, two number systems are in common use; by long tradition, natives never switch bases in mid-sentence. Suppose Alice says, "I'm a firm believer in base 10 and speak only in that base. The country has 26 people who use my base, and only 22 people speak base-14." Bob then adds, "Of the 25 residents, 13 are numerically bilingual and 1 is numerically illiterate." How many people live in Binumeria?

173. Throw Your Rings into the Hat

Imagine you have five hats arranged in a circle and 25 rings numbered 1–25. You wish to throw the rings into the hats—ring numbered 1 first, ring 2 second, and so on—so that each ring is tossed into a hat that is *adjacent* to the hat in which the preceding ring was tossed. So if 3 goes into hat A, 4 must go into hat B or E; moreover, 25 and 1 must be in adjacent hats. Can you toss all the rings into the hats in such a way that the sum of the numbers in each hat is 65?

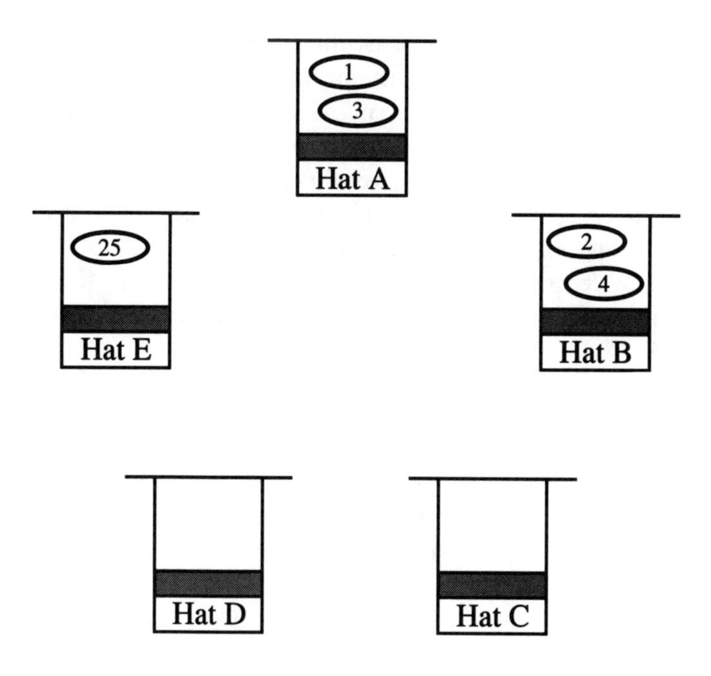

174. A Super Exponential Integral

Find, with an error of no more than 5%, the numerical value of

$$\int_1^{100} x^x \, dx.$$

175. Passion Has No Square Root

> "But you can't make arithmetic out of passion. Passion has no square root."
> —Steven Shagan (*City of Angels,* Putnam, New York 1975, p. 16)

Show that $\sqrt{\text{PASSION}} = \text{KISS}$ has a unique solution in base 10. That is, find the unique correspondence between the letters $\text{P}, \text{A}, \text{S}, \text{I}, \text{O}, \text{N}, \text{K}$ and seven of the digits $0, 1, \ldots, 9$ such that the equation holds.

176. A Problem Fit for a King

The diagram shows a pattern in which 12 distinct letters occupy a cluster of contiguous squares on a chessboard. The pattern might be called a *template* for the word "crystallographer"; that is, starting with the C and stepping from square to square as a king moves in chess, we can reach R, then Y, and so on to spell "crystallographer." Note that, as happens for the doubled L, it is legal for the move to consist of not moving. Now, if a template can contain only one copy of each letter, then for certain words it may be impossible to create a template.

<table>
<tr><td></td><td>H</td><td>E</td><td></td></tr>
<tr><td>P</td><td>A</td><td>R</td><td>C</td></tr>
<tr><td>T</td><td>Y</td><td>L</td><td>G</td></tr>
<tr><td></td><td>S</td><td>O</td><td></td></tr>
</table>

Find a string of letters (not necessarily an English word, but preferably as short a string as you can find) for which no template on a chessboard can be constructed.

177. Weigh The Boxes

A shipping clerk has five boxes of different but unknown weights. Unfortunately, all of the boxes weigh less than 100 pounds, and the only scale available reads only weights over 100 pounds. The clerk decides to weigh the boxes in pairs so that each box is weighed with every other box. The weights of all possible pairs are 110, 112, 113, 114, 115, 116, 117, 118, 120, and 121 pounds. Determine the weights of the boxes.

178. A Question of Imbalance

Five coins are identical in appearance except for labels A, B, C, D, and E. Each coin has a weight different from that of each of the others. Given an equal-arm balance, what is the minimum number of uses of the balance required to rank order the coins by weight?

179. Catch the Counterfeit

Six coins are identical in appearance, but it is possible that one of them is counterfeit. The weight of the counterfeit, if there is one, is different from that of a genuine coin, but it may be lighter or heavier. You have an accurate scale that you can use to weigh any subset of the coins. Note that this is not a balance scale, but a scale that gives a numerical reading of the weight of an object in ounces. Find a procedure for determining the weights of all the coins in only three weighings.

180. And Now For Something Completely Different

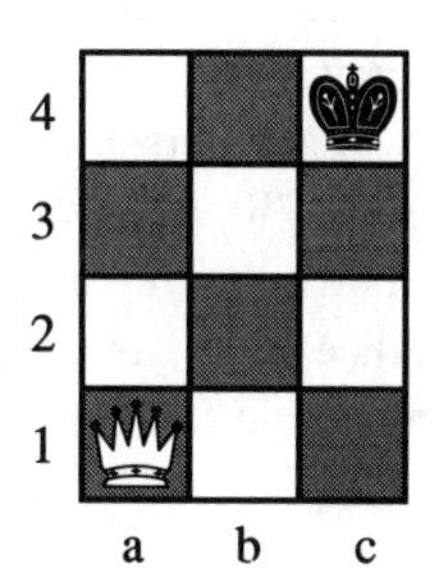

Imagine a 4×3 chessboard with a white queen at position a1 and a black king on c4. It is white's move. Assuming the rules of chess are followed (but ignoring the lack of a white king), can the white queen force the black king onto the a1 square?

181. Guess My Numbers

Alice and Bob play the following number-guessing game. Alice writes down a list of positive integers $x_1, x_2, \ldots, x_n$, but does not reveal them to Bob, who will try to determine the numbers by asking Alice questions. Bob chooses a list of positive integers $a_1, a_2, \ldots, a_n$ and asks Alice to tell him the value of $a_1x_1 + a_2x_2 + \cdots + a_nx_n$. Then Bob chooses another list of positive integers $b_1, b_2, \ldots, b_n$ and asks Alice for the value of $b_1x_1 + b_2x_2 + \cdots + b_nx_n$. Play continues in this way until Bob is able to determine Alice's numbers. How many rounds will Bob need in order to determine Alice's numbers?

182. Divide and Conquer

Alice and Bob face two boxes, one with 51 pennies in it and the other with 101. Taking turns, with Alice going first, they throw away the pennies in one box and split the contents of the other between the two boxes, in any way they choose. But for the division to be legal, it must leave at least one penny in each box. A player must make a legal move if that is possible, and the first player who cannot make a legal move is the loser. Assuming optimal strategy by both, who wins this game?

183. Divide and Be Conquered

This time Alice and Bob play the misère version of the game in Problem 182. The rules are the same except that the first player who cannot make a legal move is the *winner.* Assuming optimal strategy by both, who wins this version?

184. Greed vs. Ethics in Gambling

Alice, Bob, and Eve, all first-rate mathematicians, take part in a basketball pool based on the results of three games. Because of a point-spread system there are no draws, and the chance of guessing a game's winner correctly is assumed to be 50%. Each person puts $10 in the pot and whoever has the best prediction record gets the $30; in case of a tie, the $30 is split evenly. First Eve puts her choices into the bowl. Just before placing his choices into the bowl, Bob offers to show them to Alice (who has not yet made her choices). In return, he wants $1 from Alice. Is it profitable for Alice to agree to Bob's terms?

185. A Trigonometric Surprise

Define the sequence a_n inductively by $a_1 = 1$ and, for $n \geq 1$, $a_{n+1} = \cos(\arctan a_n)$. Find a formula for a_n.

186. Complex, Yet Simple

Let a, b, c, d denote complex numbers. True or False:

(a) If $a+b=0$ and $|a|=|b|$, then $a^2=b^2$?

(b) If $a+b+c=0$ and $|a|=|b|=|c|$, then $a^3=b^3=c^3$?

(c) If $a+b+c+d=0$ and $|a|=|b|=|c|=|d|$, then $a^4=b^4=c^4=d^4$?

187. Biased Coins

Call a biased coin a p-coin ($0 \leq p \leq 1$) if it comes up heads with probability p and tails with probability $1-p$. We say that p simulates q if by flipping a p-coin repeatedly (some finite number of times) one can simulate the behavior of a q-coin. More precisely: There exists a positive integer n and some subset of the 2^n possible outcomes of flipping the p-coin n times such that the probability of a sequence of n flips being in the subset is q.

For example, a fair coin can be used to simulate a $\frac{3}{4}$-coin by using two flips and defining a pseudo-head to be any two-flip sequence with at least one real head. The chance of a pseudo-head coming up is $\frac{3}{4}$, and so we have simulated a $\frac{3}{4}$-coin.

There is a p such that a p-coin can simulate both a $\frac{1}{2}$-coin and a $\frac{1}{3}$-coin. Find such a value.

188. An Odd Vector Sum

A nondescending vector in the plane is one whose y-component is nonnegative. The sum of two nondescending unit vectors can be very short. Prove that the sum of an odd number of nondescending unit vectors cannot have length less than 1.

189. Positive Polynomials

Suppose $p(x)$ is a polynomial of degree n and for all x, $p(x) \geq 0$. Show that for all x,

$$p(x) + p'(x) + p''(x) + \cdots + p^{(n)}(x) \geq 0.$$

190. Taming a Wild Function

The function $x \sin x$, whose graph is shown in the diagram, oscillates wildly between large positive numbers and large negative numbers as x approaches infinity. Show that there are infinitely many positive integers n such that $|n \sin n| < 4$.

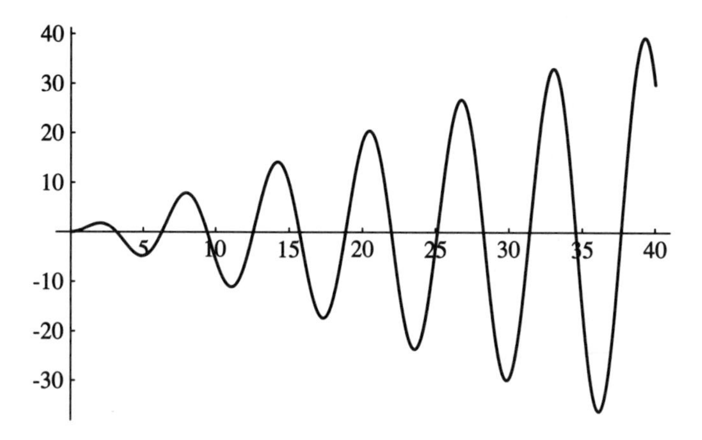

191. How Many 8s Can a Plane Hold?

A *figure eight* is a curve in the plane obtained from the basic "8" shape by any combination of translation, rotation, expansion, or shrinking; the lines forming the 8s are assumed to have no thickness. The diagram shows a bunch of figure eights in the plane. Is it possible to fit uncountably many disjoint figure 8s into the plane?

Note. A countable set is one that is finite or can be put into one-one correspondence with the integers. The set of rational numbers is countable, as is the set of points in the plane with both coordinates rational. But the set of real numbers is uncountable. Therefore the plane does contain uncountably many disjoint lines (the vertical lines, for example), or uncountably many disjoint circles (all circles centered at the origin). But the plane does not have uncountably many disjoint disks, since each disk contains a point (p, q) where p and q are rational numbers.

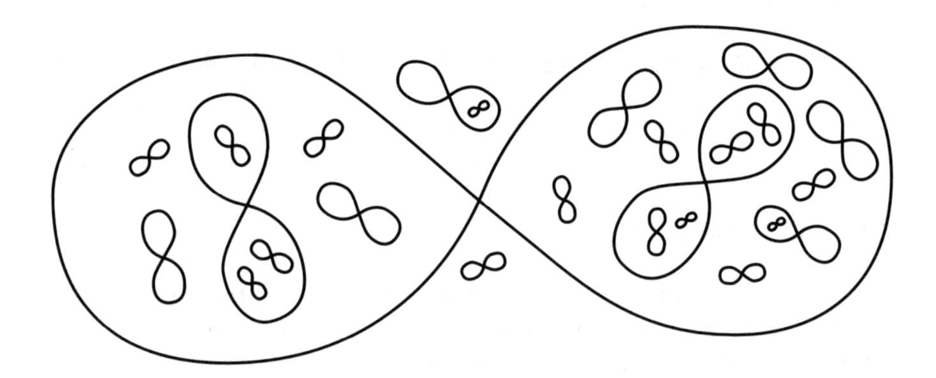

SOLUTIONS

CHAPTER 7

Plane Geometry

7.1 Locus

1. Which Way Did the Bicycle Go?

The bicyclist was going from right to left. Let $F(t)$ and $B(t)$ be the points of contact of the front and back wheels, respectively, at time t. Then, because the rear wheel does not steer, the line connecting $F(t)$ and $B(t)$ is tangent to the path of the back wheel and has *fixed length,* namely the distance between the bottoms of the bicycle wheels. It follows that the thick curve in Figure 1 cannot be the rear wheel's path, for some of its tangents fail to strike the other curve. Therefore the thin curve is the rear wheel's path.

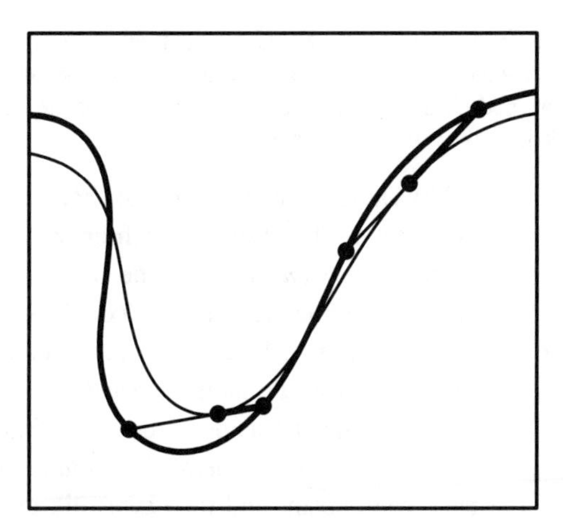

FIGURE 1
The fact that tangent segments have constant length in only one direction proves that the bicycle was traveling right to left.

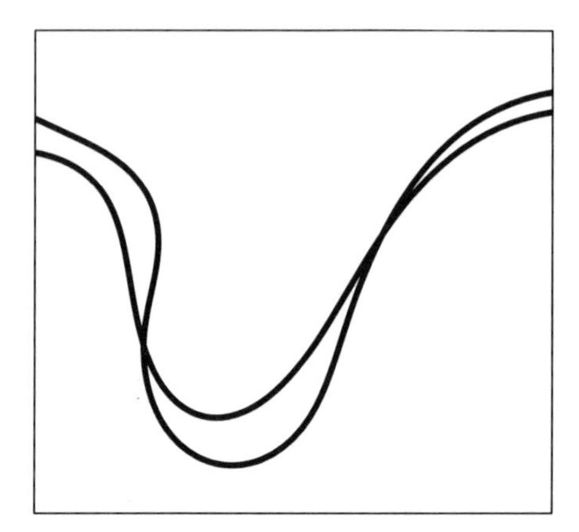

FIGURE 2
These paths show what the tracks would be if the back path were the same but the direction of travel were reversed.

We now examine some tangents to determine the direction. The lengths of the thick tangent segments in Figure 1 would be constant if the direction were from left to right. But they are not! The thin segments do have constant length, thus confirming that the direction of travel is from right to left. Figure 2 shows what the curves would look like if the back path were the same but the rider were going from left to right.

Note. Here is how Sherlock "solved" the problem:

> "No, no, my dear Watson. The more deeply sunk impression is, of course, the hind wheel, upon which the weight rests. You perceive several places where it has passed across and obliterated the more shallow mark of the front one. It was undoubtedly heading away from the school."

Balderdash! As observed by Dennis Thron (Dartmouth Medical School), it is true that the rear wheel would obliterate the track of the front wheel at the crossings, but this would be true *no matter which direction* the bicyclist was going. This information, by itself, does not solve the problem. We could, perhaps, give Holmes the benefit of the doubt and assume that he carried out the proper solution in his head. But we cannot believe that he would have expected Watson to grasp it from his comments alone. Conan Doyle has let us down badly on this one!

The bicycle paths in the figures were generated in *Mathematica* by using a Bézier curve to get the back-wheel path, and then symbolic differentiation to get the corresponding front-wheel paths in the two directions; for details see [Wag5]. We learned of this problem from materials for a geometry course at Princeton developed by John Conway, Peter Doyle, Jane Gilman, and Bill Thurston.

2. Where Can the Third Vertex Live?

If $\triangle ADX$ is equilateral and D is on BC then X is on $B'C'$ where $AB'C'$ is the triangle obtained by rotating ABC 60° in either direction. Therefore X can be on either of two line segments obtained by rotating BC 60° around A.

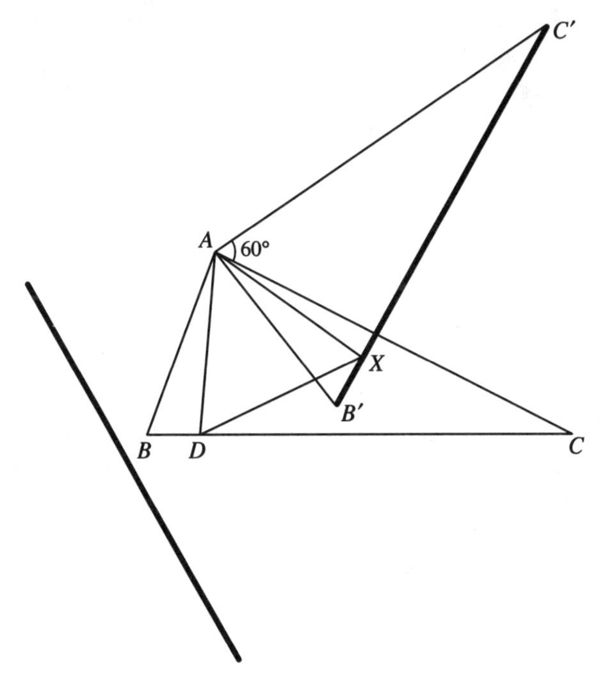

FIGURE 3

3. Seeing a 45° Angle

First observe that the shadow presented to a viewer by the square is the same as the shadow presented by either a side of the square or one of its diagonals. Now recall that if a chord in a circle subtends a 90° angle at the center, it subtends half that, 45°, at the circumference. So we need to turn the sides and diagonals into appropriate chords of circles.

Consider first the circle through B and D with center A. Chord BD makes a right angle at A, and therefore a 45° angle on the circumference. So in the region—the first quadrant of a coordinate system centered at A—for which the square's shadow is BD, this circle is part of the desired locus. Moreover, a point in this region lying inside (resp., outside) the circle sees the square at an angle greater than (resp., less than) 45°. Thus, in this first-quadrant-from-A region the

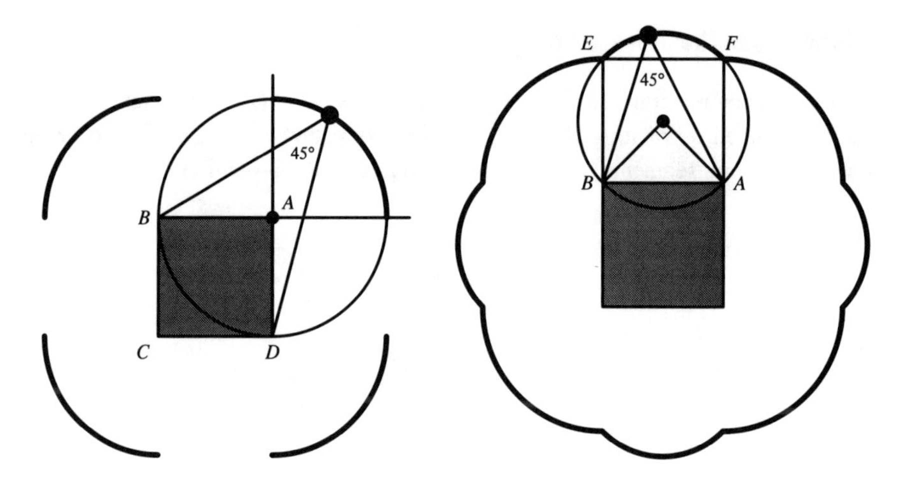

FIGURE 4

locus is precisely the quarter-circle centered at A. By symmetry, there are four such arcs in the locus. In the remaining strips, where the square's shadow is a side, use circles such as the one through A, B, E, and F in the right-hand diagram. Chord AB makes a right angle at the center of this circle, and so an arc of the circle belongs to the locus, as do four symmetrically placed arcs. The complete locus thus consists of 8 quarter-circles, four large ones and four small ones.

4. A Narrow Path

The locus is the ray bisecting $\angle UVW$. For choose B on UV and C on VW, and let A be the third vertex of the equilateral traingle ABC. Quadrilateral $ABVC$

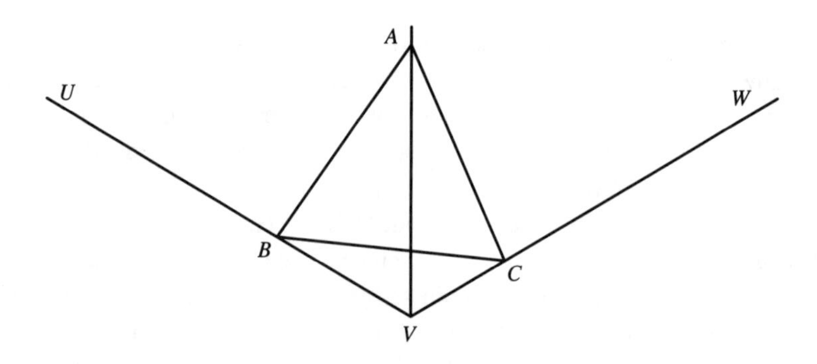

FIGURE 5

is cyclic because $\angle BAC = 60°$ and so opposite angles sum to $180°$. This means $\angle AVB = \angle ACB = 60° = \angle ABC = \angle AVC$, and therefore AV bisects the angle at V.

5. Test Your Intuition

We may assume that S has radius 1, so it is given by $x^2+y^2 = 2x$. Let C be given by $x^2 + y^2 = r^2$. Then $A = (0, r)$ and $B = \left(r^2/2, (r/2)\sqrt{4-r^2}\right)$. The slope of AB is then $\left(\sqrt{4-r^2}-2\right)/r$, and so AB intersects the x-axis at $-r$ divided by this slope, or $r^2/\left(2-\sqrt{4-r^2}\right)$. Multiplying top and bottom by the conjugate denominator yields $2+\sqrt{4-r^2}$, which approaches 4 as $r \to 0$. Thus the point X gets closer and closer to, but never reaches, the value 4.

Note. This problem is discussed in [AB], along with other similarly counterintuitive problems. It is a nice illustration of the limit concept, and appears in at least one calculus text [SB, p. 385].

6. Disks on a Circle

Choose axes so that $P = (0, 0)$ and C is the circle whose polar form is $r = \sin\theta$, $0 \le \theta < \pi$. We fix an angle θ and search for the extent of the mystery region in the θ-direction. Let Q be a typical center of one of the disks; suppose segment PQ makes an angle α with the positive x-axis. Then the radius of the Q-centered disk is $\sin\alpha$ and the corresponding circle has polar form $r = 2\sin\alpha\sin(\theta-\alpha+\pi/2)$.

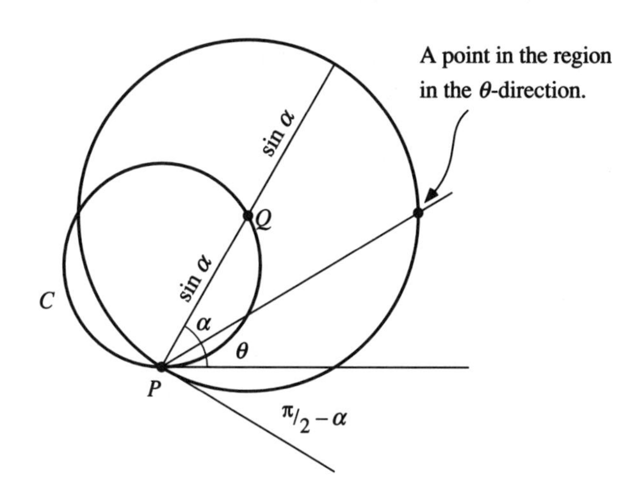

FIGURE 6

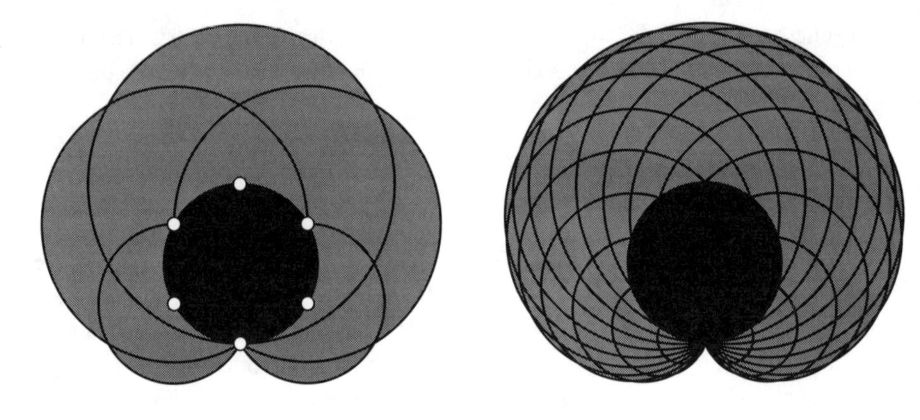

FIGURE 7

This comes from the polar form $r = 2R\sin\theta$ where R denotes radius; the radius is $\sin\alpha$ and the circle is rotated from the standard position by $\pi/2-\alpha$, which causes the shift in angle (see Figure 6).

Now, $2\sin\alpha\sin(\theta-\alpha+\pi/2) = 2\sin\alpha\cos(\alpha-\theta) = \sin(2\alpha-\theta)+\sin\theta$. Remember that θ is fixed and we seek the circle that extends farthest from P in the θ-direction. This happens when the distance from P, $\sin(2\alpha-\theta)+\sin\theta$, is maximized, which occurs for $2\alpha-\theta=\pi/2$. It follows that the desired region is bounded by the polar curve $r = 1+\sin\theta$, a cardioid.

7. Don't Cut Corners—Fold Them

Label points as illustrated and suppose the fold is along the line from R at $(0,b)$ to Q at $(a,0)$. Note that the area of the triangle is $ab/2$. Let (r,θ) be the

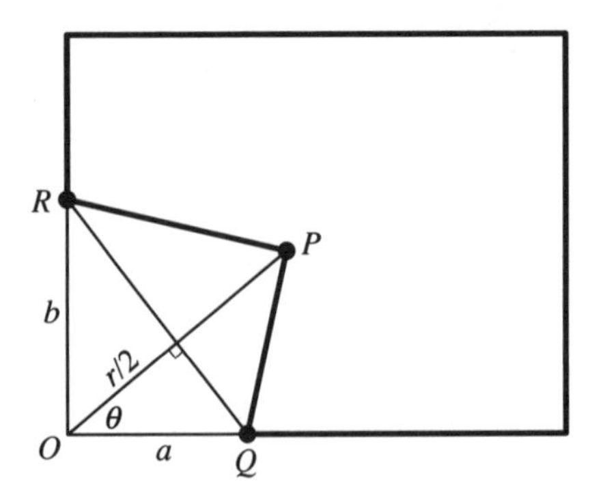

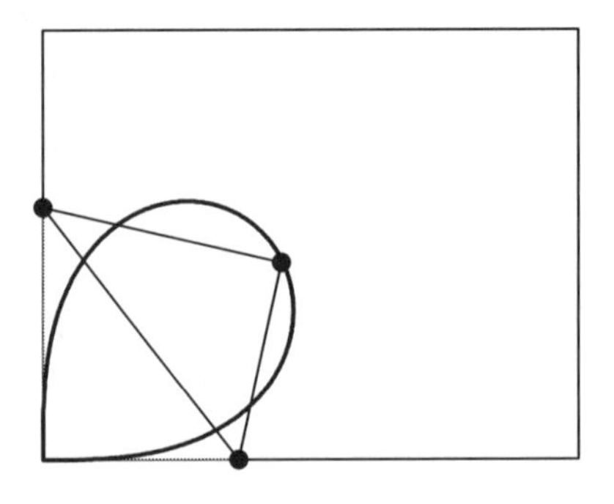

FIGURE 8

polar coordinates of P and look at the two right triangles at O, which have a common side of length $r/2$. One of them yields $r/2 = a\cos\theta$, while the other gives $r/2 = b\cos(90° - \theta) = b\sin\theta$. But by hypothesis, $ab = 2K$. Hence, $r^2/4 = ab\sin\theta\cos\theta = K\sin(2\theta)$, whence $r^2 = 4K\sin(2\theta)$, which defines a curve known as a lemniscate.

7.2 Dissection

8. Straighten these Curves

The dimensions of the rectangle in the solution in Figure 9 are $\sqrt{3} \times 3/2$.

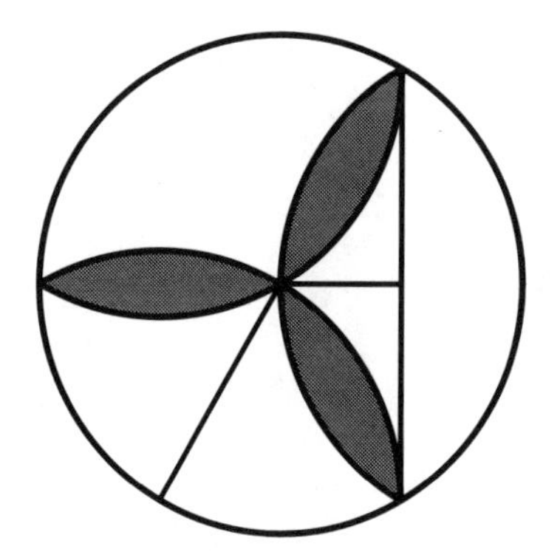

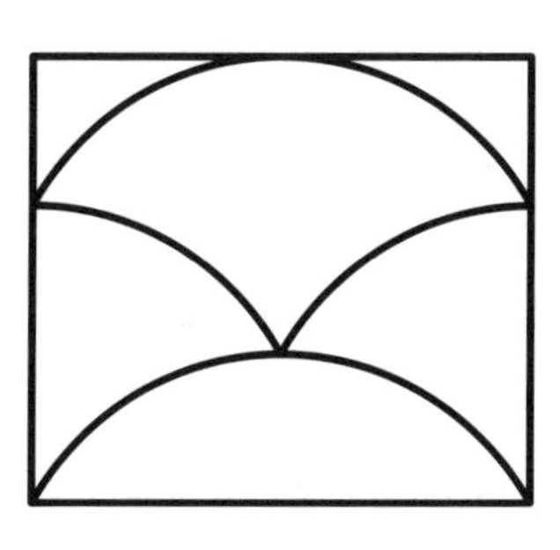

FIGURE 9

9. Cut the Triangle

Figure 10 shows the two-cut solution: just cut on the dotted lines.

Note. This problem appeared in [Eng] (see also [AKW]).

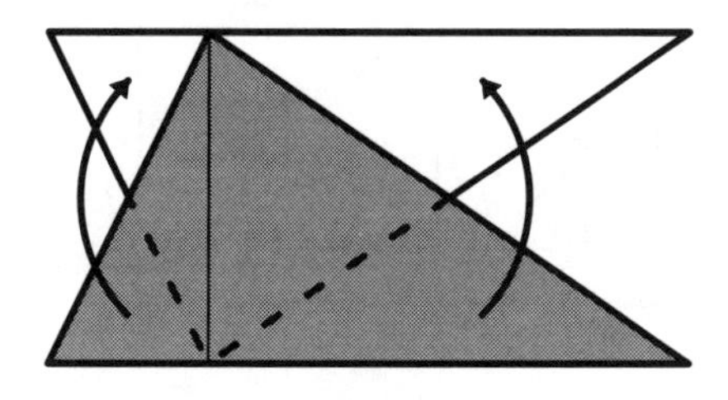

FIGURE 10

10. A Triangle Duplication Dissection

There are many possibilities; one is shown in Figure 11. All of the cuts are parallel to the sides of the triangle. When the pieces are rearranged, piece 2 is rotated 180°; the other pieces are not rotated. The cuts are made so that $AD/AB = \frac{1}{5}$, $AE/AC = \frac{3}{5}$, and $AF/AC = \frac{4}{5}$.

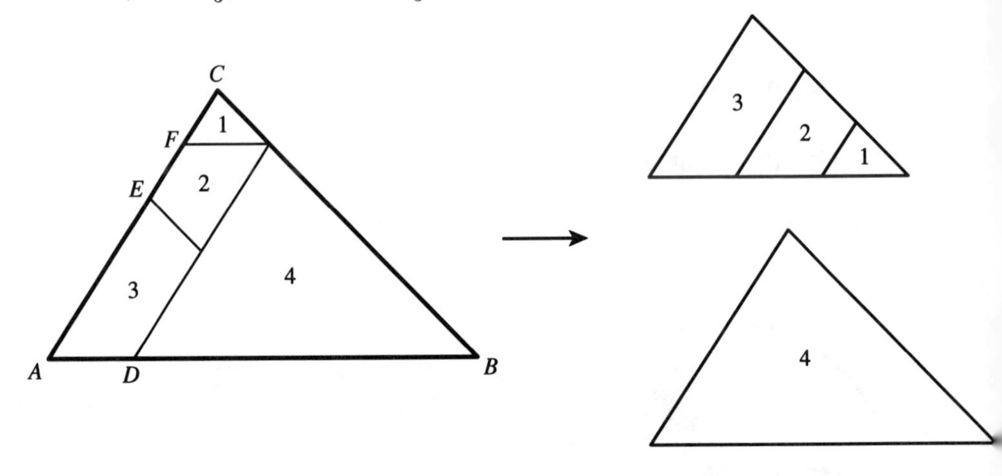

FIGURE 11

Note. This problem is due to Michael Goldberg.

11. A Square Triangulation

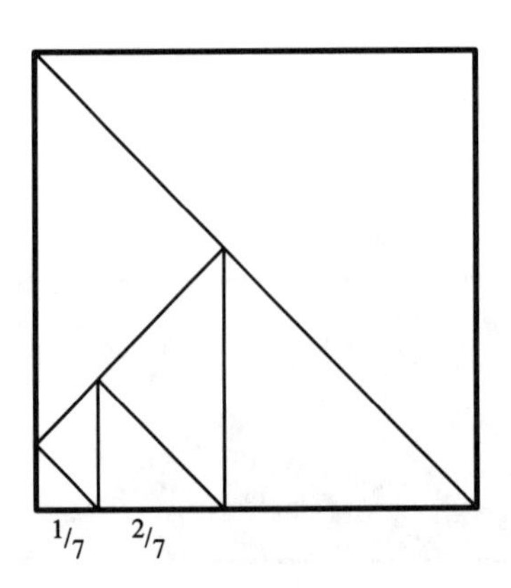

FIGURE 12

Note. This problem is due to Ivan Skvarca [Skv].

12. Equilateral Into Isosceles

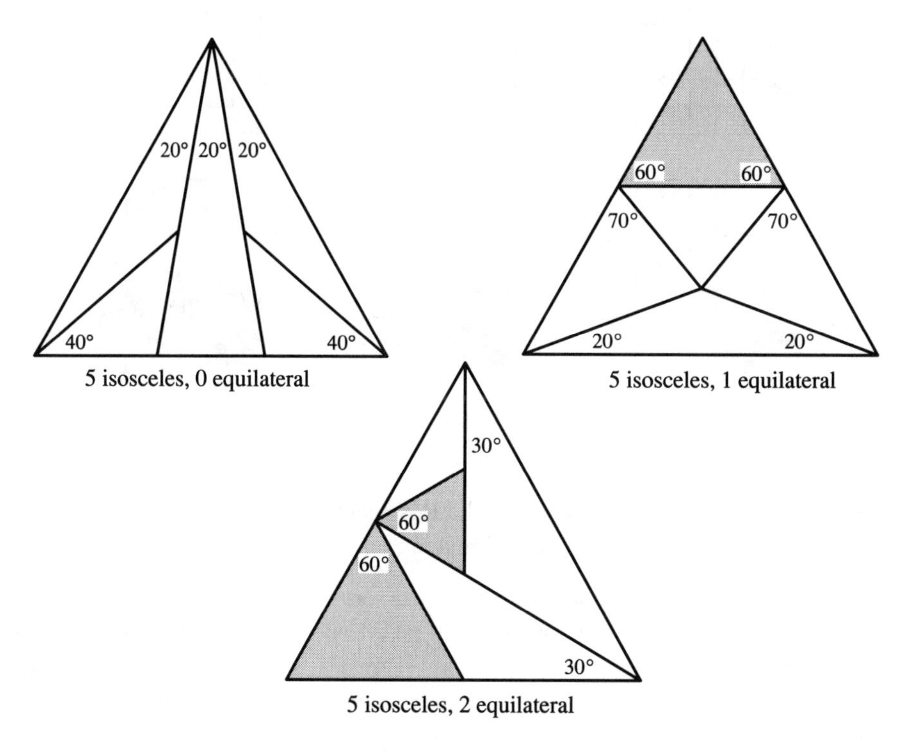

5 isosceles, 0 equilateral

5 isosceles, 1 equilateral

5 isosceles, 2 equilateral

FIGURE 13

Problem 12.1. Two of the solutions presented use 20° angles, which cannot be constructed with straightedge and compass. Constructible solutions do exist. Find them.

Problem 12.2. Can every triangle be dissected into five isosceles triangles?

Note. The question of dissecting triangles into other triangles is a fascinating one. An extensive discussion can be found in [Soi] (see also [Hon1, pp. 14–17]).

13. Solitaire on a Chessboard

Such a tiling is possible. There is essentially only one final configuration, as illustrated in Figure 14. The uniqueness result is due to the late Raphael M.

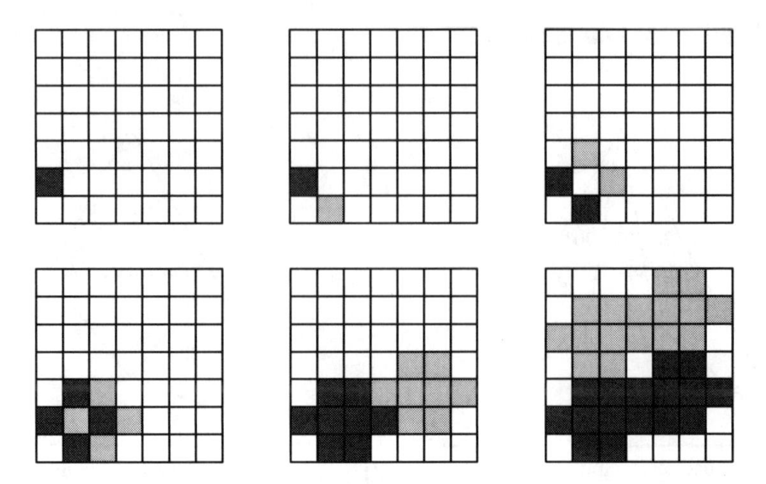

FIGURE 14

Robinson. The problem is due to Moshe Rosenfeld (Pacific Lutheran University), whose paper [Ros] contains more information on this sort of iterative tiling.

Problem 13.1 (M. Rosenfeld). Recall that a torus is obtained by identifying the opposite sides of a rectangle; thus one can go off an edge, provided one comes in at the same spot on the opposite side, as in Figure 15. In addition to the two tilings illustrated, there are other tilings of 32 squares on a 7×7 torus. How many can you find?

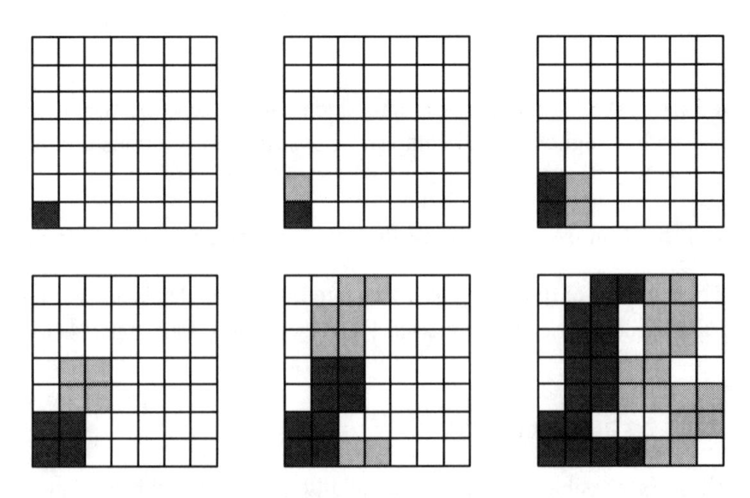

FIGURE 15

14. A Notorious Tiling Problem

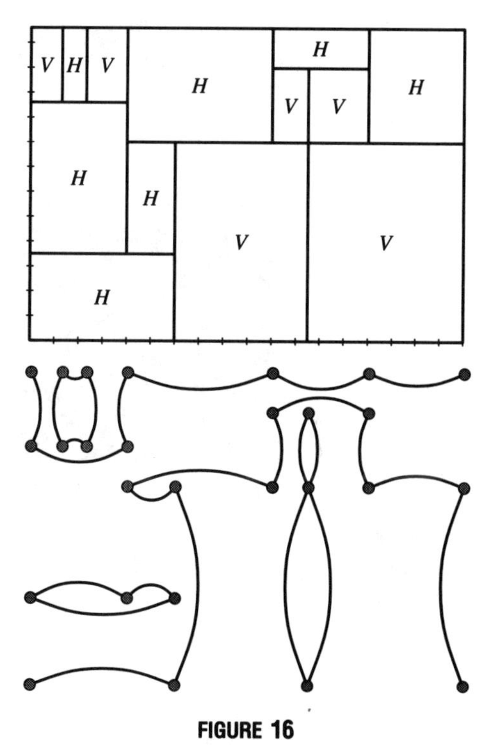

FIGURE 16

Let R denote the large rectangle; if a tile has an integer-length base, call it an H-tile (horizontal); otherwise it has integer height, and is called a V-tile (vertical). Let G be the graph whose vertices are the corners of all the tiles, with two vertices joined by an edge whenever they correspond to the ends of the horizontal sides of an H-tile or the vertical sides of a V-tile. Multiple edges may exist. Figure 16 shows a tiling and the resulting graph, which is necessarily planar. All vertices except the 4 corners of the large rectangle arise from either 2 or 4 rectangles, and hence lie on either 2 or 4 edges; the corners have only one edge.

It is well known, and not hard to prove, that a graph having some vertices of odd degree admits a path that starts at one such vertex and ends at another. In the case at hand this means there is a path that starts at one corner and finishes at another, and, because all steps correspond to integer-length distances, this proves that either the height or width (or both) of the ambient rectangle is an integer.

Notes. The graph theory proof is due to Michael Paterson, and is perhaps the simplest of the 14 proofs discussed in [Wag2]. Another quite simple proof uses the classical checkerboard coloring idea, and is closely related to de Bruijn's original proof using complex integrals. We now give this coloring proof, due independently to Richard Rochberg and Sherman Stein.

Place R in the first quadrant so that its lower-left corner is $(0,0)$. Then color the lattice generated by a $\frac{1}{2} \times \frac{1}{2}$ square in black/white checkerboard fashion. Since each tile has an integer side, each tile contains an equal amount of black and white. Therefore the same is true of R. But then R must have an integer side, for otherwise it could be split into four pieces (see Figure 17), three of which have equal amounts of black and white while the fourth does not.

Finally, we give the original proof of N. G. de Bruijn, though with his single complex integral replaced by an ordinary real double integral. First observe that

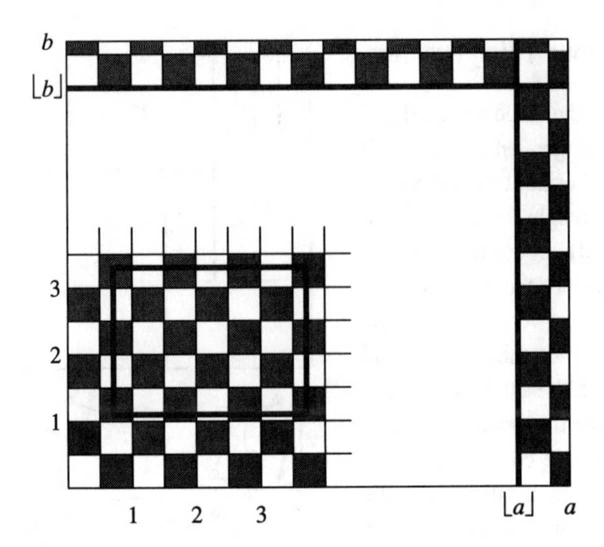

FIGURE 17
The base of the tile shown has length 3, and so black and white regions of the tile have equal area.

$$\int_a^b \sin 2\pi x\, dx = 0$$

if and only if one of $a \pm b$ is an integer. It follows that for any rectangle T whose lower-left corner is at $(0, 0)$ and whose sides are parallel to the axes,

$$\int\int_T \sin 2\pi x \sin 2\pi y\, dA = 0$$

if and only if at least one side of T has integer length. The hypothesis tells us that this double integral vanishes over each tile; therefore it vanishes over the large tile, which means the large tile has an integer-length side. The checkerboard proof is simply a discrete version of this proof, with integrand $(-1)^{\lfloor 2x \rfloor}(-1)^{\lfloor 2y \rfloor}$.

A wide variety of techniques have been brought to bear on this problem. For example, there are proofs using prime numbers, Sperner's lemma, induction, quadratic polynomials, cut-sets, sweep-lines, and step functions. See [Wag2] for a discussion of these proofs and several generalizations. Two such generalizations are:

- If a box in $\mathbb{R}^n$ is tiled with n-dimensional boxes and each tile has at least k integer sides, then the large box has at least k integer sides.

- Let G be an additive subgroup of the real numbers. If a rectangle in the plane is tiled with rectangles so that for each tile, one side has length in G, then either the base or height of the large rectangle is in G.

Problem 14.1 (S. Golomb, R. M. Robinson [Wag2]). Find a rectangle having no integer-length sides, but that can be tiled toroidally with rectangles each of which has at least one integer side. The notion of a toroidal tiling is explained in Problem 13.1.

Finally, we note that Richard Kenyon [Ken] has recently come up with a totally new and simple proof using multiplication in a free product of two groups ($S^1 * S^1$).

7.3 Triangles

15. An 80°-80°-20° Triangle

Construct the figure illustrated using three 80°-80°-20° triangles, and let Q be on AE so that $\angle APQ = 60°$. Then $\triangle APQ$ is equilateral and $PQ = AP = CD$. Congruent triangles yield $CP = DQ$, so quadrilateral $PQDC$ has opposite sides equal, and is therefore a parallelogram. But by the figure's symmetry, each of the four angles of quadrilateral $PQDC$ is a right angle (easily proved by congruent triangles). Therefore $\angle ACP = 90° - \angle ACD = 90° - 80° = 10°$.

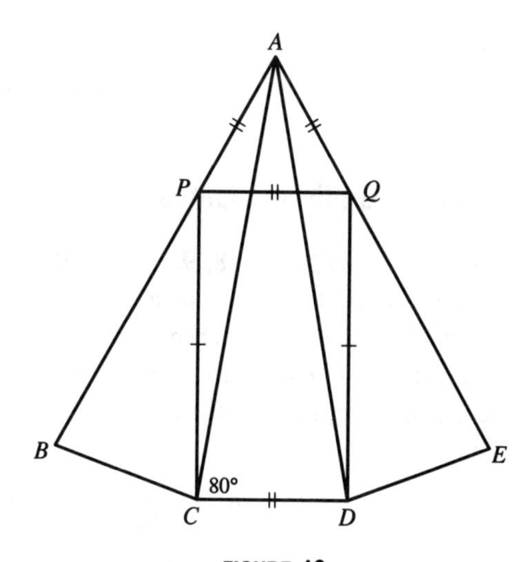

FIGURE 18

Notes. This problem is a classic; it appeared in [JD]. For another famous problem involving an 80°-80°-20° triangle see [Hon, pp. 16–18].

16. A Decomposition of Unity

Let s and t be the lengths marked in Figure 19. Then since all the triangles in the figure are similar, $b'/b = s/c$ and $a'/a = t/c$. Therefore

$$\frac{a'}{a} + \frac{b'}{b} + \frac{c'}{c} = \frac{t}{c} + \frac{s}{c} + \frac{c'}{c} = \frac{t+s+c'}{c} = \frac{c}{c} = 1.$$

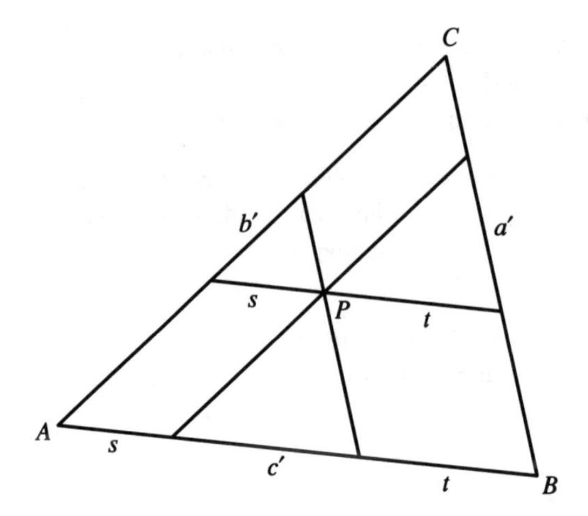

FIGURE 19

17. Find the Missing Altitude

The possibilities are: 7, 8, 9, 10, 11, 12, or 13. To prove this, let the lengths of the sides of the triangle be a, b, and c, and let the altitudes to these sides be h_a, h_b, and h_c, respectively. Then $ah_a = bh_b = ch_c$, since each is twice the triangle's area; thus $a : b : c = h_b : h_a : h_a h_b / h_c$ and $h_a : h_b : h_c = b : a : ab/c$. Applying the triangle inequality to the first of these equations we must have

$$h_b < h_a + \frac{h_a h_b}{h_c}, \quad \text{or} \quad \frac{1}{h_c} > \frac{h_b - h_a}{h_a h_b},$$

$$h_a < h_b + \frac{h_a h_b}{h_c}, \quad \text{or} \quad \frac{1}{h_c} > \frac{h_a - h_b}{h_a h_b}, \quad \text{and}$$

$$\frac{h_a h_b}{h_c} < h_a + h_b, \quad \text{or} \quad \frac{1}{h_c} < \frac{h_a + h_b}{h_a h_b}.$$

In other words,

$$\frac{|h_a - h_b|}{h_a h_b} < \frac{1}{h_c} < \frac{h_a + h_b}{h_a h_b}.$$

Since the triangle inequality is the only restriction on the lengths of the sides of a triangle, if h_a, h_b, and h_c are any three positive numbers satisfying this last inequality then a triangle with side lengths in the ratio $a : b : c = h_b : h_a : h_a h_b / h_c$ can be constructed. The ratio of the altitudes in this triangle would then be $b : a : ab/c = h_a : h_b : h_c$, so a triangle similar to this one would have altitudes

h_a, h_b, and h_c. Thus the inequality we have derived is the only restriction on the altitudes of a triangle.

In our case we have $h_a = 9$ and $h_b = 29$, so $\frac{20}{261} < 1/h_c < \frac{38}{261}$ and $6.86 < h_c < 13.05$.

Problem 17.1. Do there exist positive integers h_1 and h_2 such that there does not exist a triangle with three integer altitudes, two of which are h_1 and h_2?

Problem 17.2. Do there exist two distinct integers h_1 and h_2 such that the only triangle with three integer altitudes, two of which are h_1 and h_2, is isosceles?

18. An Isosceles Chain

For an isosceles triangle T, let $E(T)$, the *equilateral discrepancy,* denote $60° - B$, where B is the base angle. Then it is easy to verify that $E\big(K(T)\big) = -\frac{1}{2}\big(E(T)\big)$. Because a base angle is between $0°$ and $90°$, $E(T)$ lies between $60°$ and $-30°$. Moreover, the angles of an isosceles triangle all have an integer number of degrees if and only if $E(T)$ is an integer number of degrees. Now, the longest possible integer sequence of the form $m, \frac{m}{2}, \frac{m}{4}, \frac{m}{8}, \ldots$ occurs when m is as high a power of 2 as possible. Since 32 is the highest power of 2 between -30 and 60, the longest such sequence has length 6 and is $32, -16, 8, -4, 2, -1$. The base angles of the longest chain of triangles are therefore: $28°$, $76°$, $52°$, $64°$, $58°$, $61°$.

Note. This problem appeared in [Sil3].

19. Nested Triangles

No. Let the vertices of the original triangle be A, B, and C, and let the measures of the angles at these vertices be α, β, and γ respectively. Let P be the center of the inscribed circle, and let D, E, and F be the points of tangency, as in Figure 20. Then since $\angle ADP$ and $\angle AFP$ are right angles and the angles of the quadrilateral $ADPF$ sum to $360°$, we must have $\angle DPF = 180° - \alpha$. Since $\triangle DPF$ is isosceles, it follows that $\angle FDP = \alpha/2$. Similarly, $\angle EDP = \beta/2$, so $\angle EDF = (\alpha + \beta)/2$. Similar reasoning shows that $\angle DEF = (\beta + \gamma)/2$ and $\angle DFE = (\alpha + \gamma)/2$. In other words, the angles of $\triangle DEF$ are the averages of the angles of $\triangle ABC$ taken two at a time.

Suppose without loss of generality that $\alpha \leq \beta \leq \gamma$. Then the difference between the largest and smallest angles of $\triangle ABC$ is $\gamma - \alpha$, and since we have assumed the triangle is not equilateral, this is nonzero. The largest and smallest angles of $\triangle DEF$ are $(\beta + \gamma)/2$ and $(\alpha + \beta)/2$ respectively, and the difference between

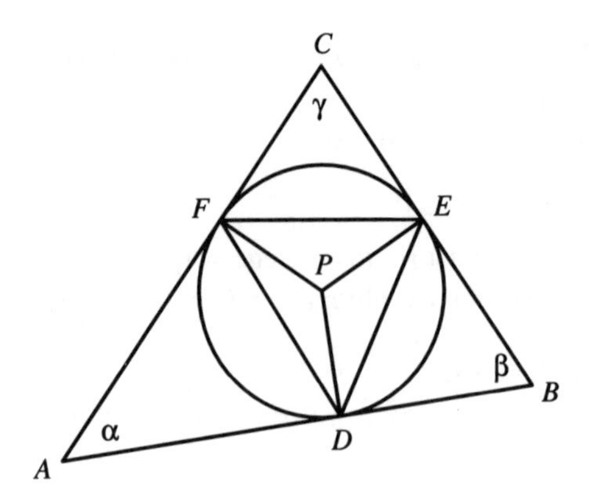

FIGURE 20

these is $(\gamma - \alpha)/2$. Thus, the difference between the largest and smallest angles of $\triangle DEF$ is half as big as the corresponding difference in $\triangle ABC$. Clearly if the process of nesting triangles is continued, then at each step the difference between the largest and smallest angles will be half as big as it was at the previous step. Thus none of the triangles can be similar.

20. Avoiding Equilaterals

No. In fact, even the ten numbered points in the grid of Figure 21 cannot be divided into two sets so that neither contains the vertices of an equilateral triangle.

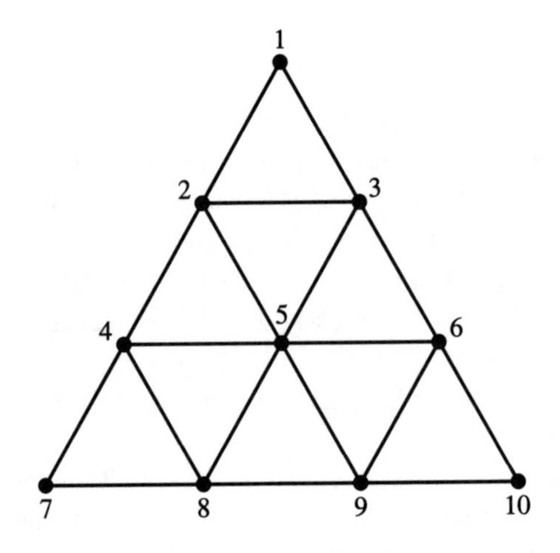

FIGURE 21

To see why, suppose the ten points in the grid have been divided into two sets A and B, neither of which contains the vertices of an equilateral triangle. Without loss of generality assume that point 5 is in set A. Since points 3, 4, and 9 form an equilateral triangle, at least one of them must be in A. Assume without loss of generality that it is 4. Then clearly 2 and 8 must be in B, since either would form an equilateral triangle with 4 and 5. Since 2, 8, and 6 form an equilateral triangle, 6 must be in A, so 3 must be in B. But now if 1 is in A then A contains the equilateral triangle 1, 4, 6, and if it is in B then B contains the equilateral triangle 1, 2, 3.

21. The Square on the Hypotenuse

Let the sides of the large triangle be a, b, and c, which we know to be integers; $a^2 + b^2 = c^2$. Let x be the side-length of the square. The hypotenuse of the big triangle comes in three pieces: the central one has length x; let y and z denote the lengths to the left and right. The small triangles to the left and right of the square are similar to the large triangle, so we can solve for y and z to get $y = ax/b$ and $z = bx/a$. Thus $c = ax/b + x + bx/a = x(a^2 + ab + b^2)/(ab)$, whence $x = abc/(a^2 + ab + b^2) = abc/(c^2 + ab)$. We are interested in Pythagorean triples (a, b, c) for which this is an integer, so we want $c^2 + ab$ to divide abc.

Now, write $(a, b, c) = (km, kn, kp)$, with m, n, and p pairwise relatively prime. Substituting yields $x = kmnp/(p^2 + mn)$, so we need $p^2 + mn$ to divide $kmnp$. But $p^2 + mn$ is relatively prime to m, n, and p, so we must have $(p^2 + mn)|k$; say $k = (p^2 + mn)t$. This means $x = mnpt$. Note that this characterizes exactly all integer right triangles with integer inscribed squares: they are precisely those triangles with sides (km, kn, kp) where (m, n, p) is a primitive Pythagorean triple and k is a multiple of $p^2 + mn$.

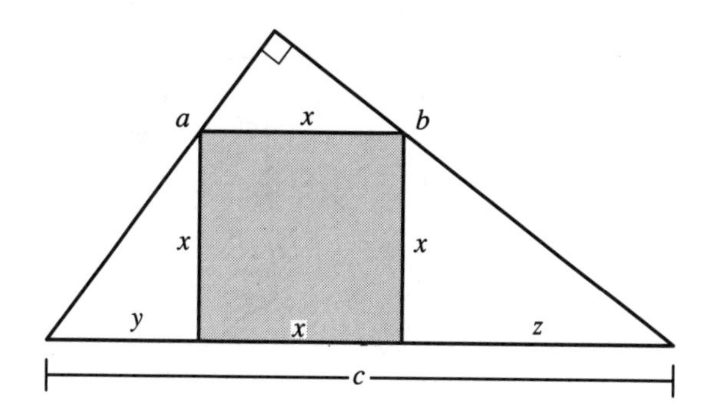

FIGURE 22

Returning to our problem, we want to minimize x, so obviously we take $t = 1$ to get $x = mnp$. We now seek the primitive Pythagorean triple (m, n, p) for which mnp is minimal; this is $(3, 4, 5)$, with $x = 3 \cdot 4 \cdot 5 = 60$. Then $k = p^2 + mn = 25 + 12 = 37$ and the original triangle has dimensions $37(3, 4, 5) = (111, 148, 185)$.

22. A Hexagon-Triangle Hinge

Extend the figure by reflection in M as illustrated. Then M is the midpoint of QQ'. We first show that triangles PCQ and $P'BQ$ are congruent. This can be done by side-angle-side as follows. Two sides are clear. Let the angles of $\triangle ABC$ be α, β, γ at vertices A, B, C, respectively. Then $\angle PCQ = 360° - 90° - \gamma = 270° - (180° - \alpha - \beta) = 90° + \alpha + \beta = \angle QBP'$; this gives the angle of the side-angle-side. This yields $PQ = QP'$, and so $PQP'Q'$ is a rhombus. Since the diagonals of a rhombus bisect each other at right angles, $\angle PMQ = 90°$.

Note. This problem is due to Stanley Rabinowitz [Rab1] and Jack Garfunkel [Gar]. The proof given above yields a generalization found by Mary Krimmel [Kri]. Let P be the apex of an isosceles triangle erected outward on side AC of $\triangle ABC$, let Q be the apex of an isosceles triangle erected outward on side BC of $\triangle ABC$, and let R be the midpoint of AB. If the angles at P and Q sum to $180°$, then $\angle PRQ = 90°$.

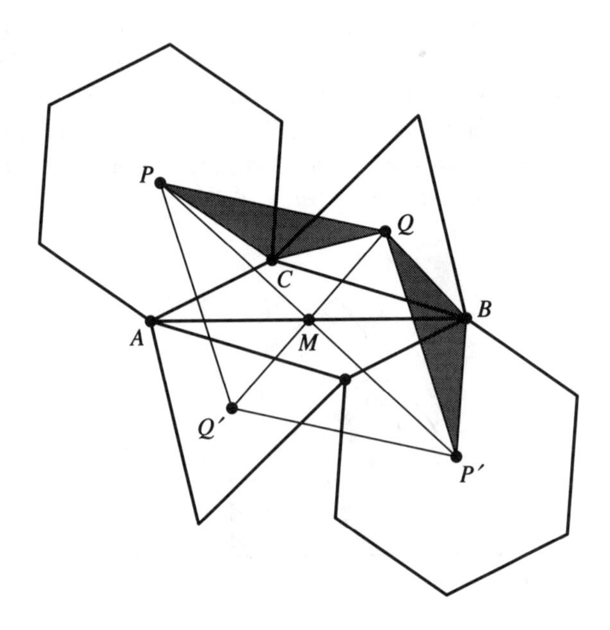

FIGURE 23

23. How Many Isosceles Triangles?

If a point makes an isosceles triangle with AB, then it is either on the perpendicular bisector of AB or on one of the two circles with radius AB and centered at A or B. Figure 24 shows lines meeting this point-set in 0, 1, 2, 3, 4, or 5 points, and no line can meet the set in more than 5 points. Thus the possible finite values of $n(L)$ are 0, 1, 2, 3, 4, or 5.

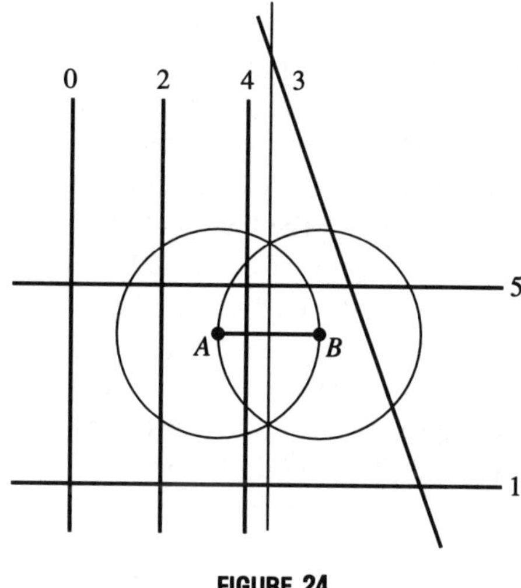

FIGURE 24

24. A Triangle Bisection

In Figure 25, CE bisects $\angle ACB$. We assume without loss of generality that N is on AC. Extend AC to a point D such that $CD = CB$, and draw the segment BD. Then $\triangle BCD$ is isosceles, so $\angle CDB = \frac{1}{2}(180° - \angle DCB) = \frac{1}{2}\angle ACB = \angle ACE$. Therefore BD is parallel to CE and MN, so $\triangle AMN$ and $\triangle ABD$ are similar. Since M is the midpoint of AB, we have $AB = 2AM$ and $AD = 2AN$. Thus the perimeter of $\triangle ABC$ is $AB + AC + CB = AB + AC + CD = AB + AD = 2(AM + AN)$, as required.

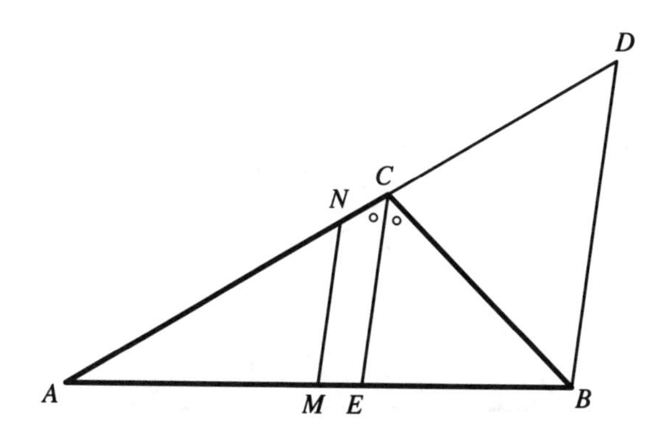

FIGURE 25

25. A Cut Through the Centroid

Divide each side of the triangle into three equal pieces and connect the division points by segments parallel to the sides of the triangle, dividing the triangle into nine congruent triangles, each having area $\frac{1}{9}$. Note that the point O in the center of this grid is the centroid of the triangle.

It is convenient to consider first the lines through O that cut the triangle most and least evenly. The line through C and O cuts the side AB at its midpoint D. Thus the triangles ADC and DBC have the same base and altitude, so they both have area $\frac{1}{2}$. At the other extreme, it is clear that UV cuts the triangle into pieces with areas $\frac{5}{9}$ and $\frac{4}{9}$. We will show that every other line through O cuts the triangle into pieces whose areas fall somewhere between these extremes.

Consider the line PQ. Note that triangles QDO and RSO are congruent, as are triangles VQO and URO. Thus, writing $\alpha(X)$ for the area of a region X, the area of the piece of the triangle to the right of PQ is

$$\begin{aligned}\alpha(QBP) &= \alpha(DBC) + \alpha(QDO) - \alpha(PCO)\\ &< \alpha(DBC) + \alpha(QDO) - \alpha(RSO)\\ &= \alpha(DBC) = \frac{1}{2}.\end{aligned}$$

Similarly,

$$\begin{aligned}\alpha(QBP) &= \alpha(VBU) + \alpha(UPO) - \alpha(VQO)\\ &> \alpha(VBU) + \alpha(URO) - \alpha(VQO)\\ &= \alpha(VBU) = \frac{4}{9}.\end{aligned}$$

Similar reasoning can be applied to any line through O to establish the desired conclusion.

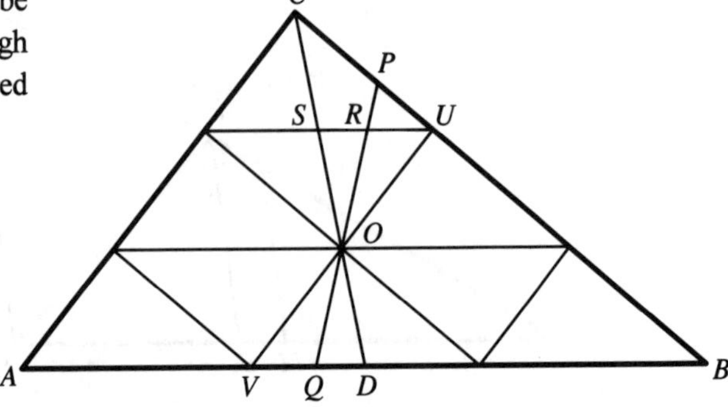

FIGURE 26

26. A Common Ratio

The desired ratio is the golden ratio. If E is the other intersection point of the circle with the line determined by MN, then triangles AEN and CFN are similar (the angles at C and E are equal, since they cut off the chord AF). Therefore $NE/NA = NC/NF$. But $\triangle AMN$ is equilateral, so $NA = MN$. Also, $MF = NE$ and $NC = MN$. Therefore $MF/MN = MN/NF$.

For the rest, let t be the ratio of MN to NF. Then we have

$$t = \frac{MN}{NF} = \frac{MF}{MN} = \frac{MN + NF}{MN}$$

$$= 1 + \frac{NF}{MN} = 1 + \frac{1}{t}.$$

Thus $t^2 = t + 1$, and t is therefore the golden ratio, $\frac{1}{2}(1 + \sqrt{5})$.

Note. This problem is due to George Odom [Odo].

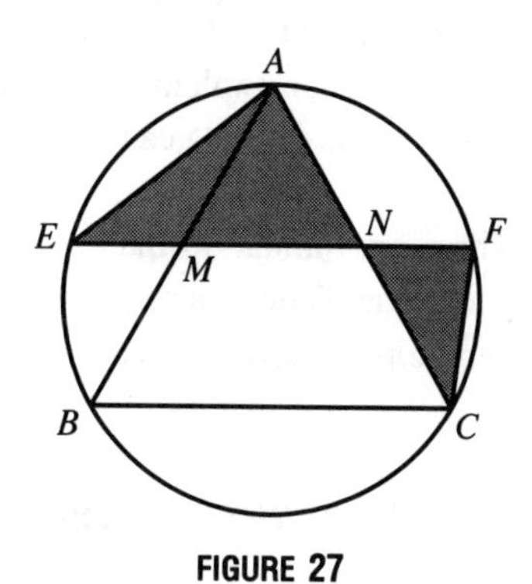

FIGURE 27

27. A 3-4-5 Triangle Problem

On AC construct $\triangle AQC$ congruent to $\triangle APB$ with Q outside $\triangle ABC$. Then $\angle PAQ = \angle BAC = 60°$; it follows that $\triangle APQ$ is equilateral with side-length 3. Since $PC = 5$ and $QC = 4$, this means that $\angle PQC$ is a right angle, and so $\angle AQC = 150°$. The law of cosines on $\triangle AQC$ then tells us that

$$AC = \sqrt{9 + 16 - 24\cos 150°}$$

$$= \sqrt{25 + 12\sqrt{3}}.$$

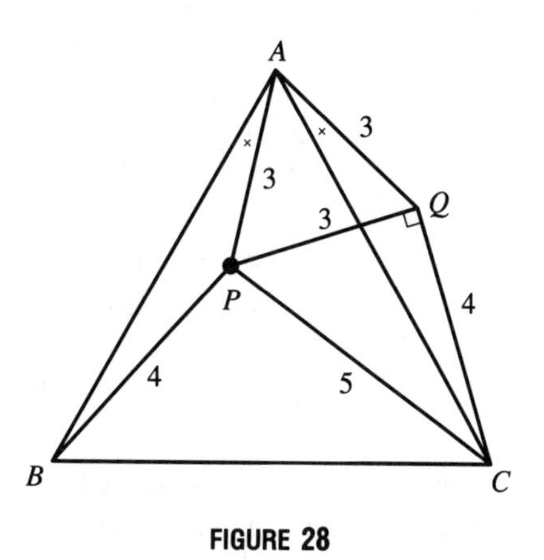

FIGURE 28

Problem 27.1 (L.-S. Hahn). Triangle ABC is equilateral and P is in its interior. The distances PA, PB are 3, 4, respectively. Assuming the edge length of the triangle is 5, what is the distance PC?

Notes. This puzzle is a famous one and has a long history. Less well known is the fact that there is a general formula for this sort of problem. For the case of a triangle and a point P in the same plane, with a_i being the distances from P to the three vertices and a_0 being the side-length of the triangle, we have:

$$a_1^2 a_2^2 + a_2^2 a_3^2 + a_1^2 a_3^2 - \sum_{j=1}^{3} a_j^4 + a_0^2 \sum_{j=1}^{3} a_j^2 = a_0^4.$$

This is a quadratic in a_0^2 and so easily yields a_0. Note that the formula is equivalent to the elegant form: $(\sum_{j=0}^{3} a_j^2)^2 = 3\sum_{j=0}^{3} a_j^4$. This version reveals an interesting symmetry, which explains why the answers to Problems 27 and 27.1 are identical! A proof of this formula can be constructed along the lines of the solution to Problem 27.

There are similar formulas in the case of the analogous problem for a regular n-gon in the plane or a regular $(n+1)$-simplex in n-dimensional space. See [Rab5] for a comprehensive discussion of the circle of ideas related to this problem.

28. Triangles from a Triangle

We use vectors. The sides of the triangle we seek will correspond to $\pm(X - A)$, $\pm(X-B)$, and $\pm(X-C)$, and so X must satisfy $\pm(X-A)\pm(X-B)\pm(X-C) = 0$. There are eight possibilities, but only four distinct points.

$$\begin{array}{ll}
+(X-A)+(X-B)+(X-C)=0 & X=(A+B+C)/3 \\
+(X-A)+(X-B)-(X-C)=0 & X=A+B-C \\
+(X-A)-(X-B)+(X-C)=0 & X=A-B+C \\
+(X-A)-(X-B)-(X-C)=0 & X=-A+B+C \\
-(X-A)+(X-B)+(X-C)=0 & X=-A+B+C \\
-(X-A)+(X-B)-(X-C)=0 & X=A-B+C \\
-(X-A)-(X-B)+(X-C)=0 & X=A+B-C \\
-(X-A)-(X-B)-(X-C)=0 & X=(A+B+C)/3
\end{array}$$

Geometrically these points are the centroid of the triangle and the three points that form a parallelogram with two of the triangle's sides; equivalently: the centroid and the three points obtained by reflecting each vertex in the midpoint of the opposite side.

Problem 28.1. Suppose P is an n-sided polygon in the plane. How many points X exist such that the segments XV, V a vertex of P, can be translated to form an n-gon? Hint: There is a simple formula, with two cases according to the parity of n.

Problem 28.2. Suppose $ABCD$ is a quadrilateral in the plane. Find a geometric description of the set of points X such that the segments XA, XB, XC, and XD can be translated to form a quadrilateral. What about the same problem for a pentagon?

29. An Equilateral Triangle from a Circle and Hyperbola

Use polar coordinates with origin at the center of the circle. Suppose the point (x, y) is on both the circle and the hyperbola; then (x, y) has polar coordinates (r, θ) where $r^2 = a^2 + b^2$ and

$$r^2 \cos^2 \theta - r^2 \sin^2 \theta + ar \cos \theta + br \sin \theta = 0$$

But the point (a, b) is itself on the circle and so $(a, b) = r(\cos \phi, \sin \phi)$ for some angle ϕ. The hyperbola equation thus becomes:

$$\begin{gathered} r^2 \cos^2 \theta - r^2 \sin^2 \theta + r^2 \cos \phi \cos \theta + r^2 \sin \theta \sin \phi = 0, \quad \text{or} \\ \cos^2 \theta - \sin^2 \theta + \cos \phi \cos \theta + \sin \theta \sin \phi = 0, \quad \text{or simply} \\ \cos(2\theta) + \cos(\theta - \phi) = 0. \end{gathered}$$

Hence 2θ is either $\theta - \phi + \pi + 2n\pi$ or $\phi - \theta + \pi + 2n\pi$, for some integer n. In the first case, $\theta = -\phi + (2n + 1)\pi$, which yields one of the four points, while in the second case, $\theta = (\phi + \pi)/3 + 2n\pi/3$, for the other three points. But these last three points are equally spaced around the circumference, and so form an equilateral triangle.

Note. This problem is due to Norman Anning [Ann].

7.4 Circles

30. Abe Lincoln's Somersaults

It's best to first analyze the case of a single penny rolling all the way around another one. In that case the rolling penny makes *two* revolutions about its center (get out some pennies—quarters are better; after all George Washington was a revolutionary president—and try it!). This is because a roll that unravels θ worth of circumference can be decomposed into a sliding part (no rolling) and a spinning part (where the rolling is taken into account); each part contributes θ to the answer, so the total rotation is 2θ. Referring to Figure 29, note that during the sliding part, the wheel's radius to P would become the radius emanating to Q. But during the spinning part there would be an additional rotation to R through the angle θ.

The problem posed involves a rolling through an angle of $6(2\pi/3)$, which is equivalent to two full circumferences. As the penny rolls, it "thinks" it is rolling around a single penny, and so the 2θ rule applies; thus the circuit is equivalent to a penny rolling twice around another one, yielding the answer of twice two, or four.

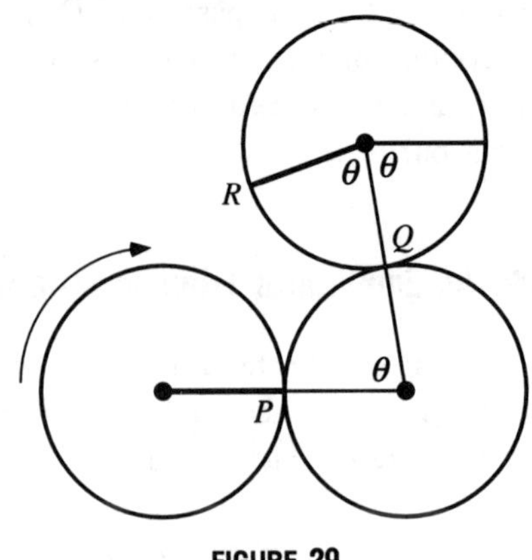

FIGURE 29

31. Focus on This

Let the radius of the bicycle wheel be r; let O be the center of the wheel, let P be an arbitrary point on the rolling disk, and let d be the distance from O to P. A key idea in rolling wheel problems is to think of the rolling as being made up of a rightward slide (with no rolling) followed by a spin about the center. We can therefore think of the vector describing the velocity of a point P on the disk as the

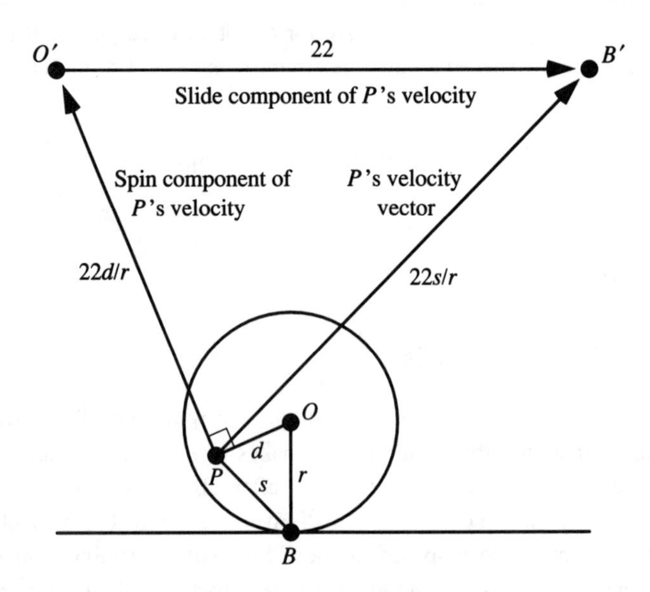

FIGURE 30

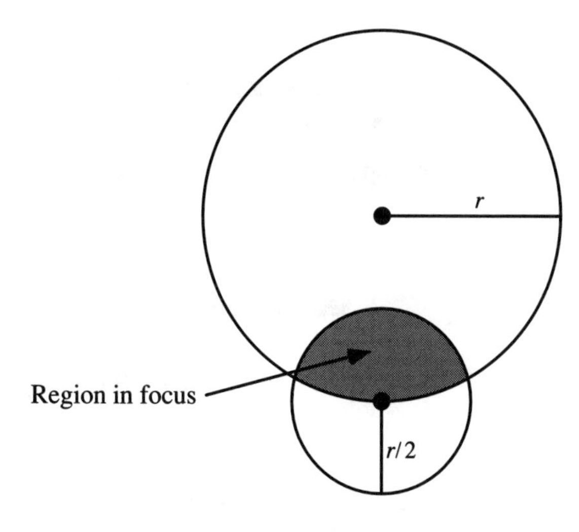

FIGURE 31

vector sum of a sliding and a spinning component. For more precision, observe that the wheel's circumference unravels 22 feet in one second; therefore the wheel spins through $22/(2\pi r)$ revolutions in a second and P spins through $2\pi d \cdot 22/(2\pi r)$, or $22d/r$ feet per second. Thus the spinning component of P's velocity is the vector that is perpendicular to OP and has length $22d/r$. The sliding component is simply a horizontal vector of length 22. In fact, there is a simple geometric construction that computes the sum of these two vectors.

Let B be the bottom point of the circle and let s be the distance from P to B. Form the triangle OPB and then rotate it 90° counterclockwise around the point P, simultaneously enlarging it by a factor of $22/r$. Let B' and O' be the points that arise from applying this transformation to B and O, respectively. Then $\triangle B'O'P$ has sides 22, $22d/r$, and $22s/r$.

The vector $O'B'$ is horizontal and has length 22, so it represents the sliding component of P's velocity. And PO' has length $22d/r$ and is perpendicular to OP, so it is the spinning component. Therefore the vector sum, PB', is the overall velocity vector of P. But this vector has length $22s/r$, which is therefore P's speed. This tells us that the speed depends only on P's distance from B. It then follows that the answer to the question is: All points in the disk for which $22s/r \leq 11$, or simply all points on the wheel that lie in a disk of radius $r/2$ centered at B. The intuitive interpretation of what has been done here is that the instantaneous motion of P as the wheel rolls is the same as if P were being rotated along a circle centered at B.

Figure 32 shows how it would look to a stationary camera.

FIGURE 32

Note. Variations of this problem appear in [Bol] and [dSV].

32. Circular Surprises

Let the desired radii be a and b in each case, with $a < b < 1$, and let O, P, Q, and R be as in Figure 33, which is for the specific case of the triangle but can be used as a guide to all three cases. The right-angled triangles PRQ and OQR yield

$$\begin{aligned} PQ^2 &= PR^2 + QR^2 \\ (a+b)^2 &= (b-a)^2 + OQ^2 - OR^2 \\ 4ab &= (1-a)^2 - \big(1-b-(b-a)\big)^2 \\ 4ab &= 1 - 2a + a^2 - (1-2b+a)^2 \\ 4ab &= -4a + 4b + 4ab - 4b^2 \\ a &= b(1-b). \end{aligned}$$

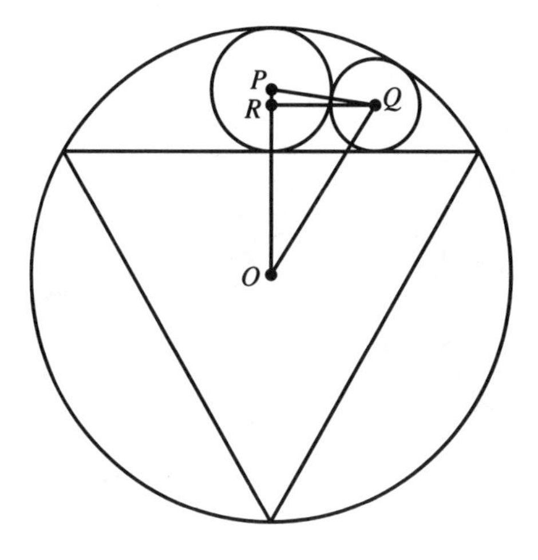

FIGURE 33

In the three cases b is easily determined, yielding the following values:

$$\begin{array}{ll} b = \frac{1}{2} & a = \frac{4}{16} \\ b = \frac{1}{4} & a = \frac{3}{16} \\ b = \frac{1}{4}(2-\sqrt{2}) & a = \frac{2}{16} \end{array}$$

Note that the b-values may be interpreted as:

$$\frac{2-\sqrt{0}}{4}, \frac{2-\sqrt{1}}{4}, \frac{2-\sqrt{2}}{4}.$$

Problem 32.1. Show that the patterns continue if one jumps from a square to an inscribed hexagon.

Problem 32.2. What are the radii in the case of a pentagon?

Note. This pretty pattern is due to Leon Bankoff [Ban], a mathematician and dentist from Los Angeles who has posed many interesting geometrical problems. The cover of the March 1992 issue of the *College Mathematics Journal* [Ale] shows Bankoff doing dentistry work on a leopard.

33. Tangent Cuts

Let the radii of the circles centered at A and B be r_A and r_B, respectively, and let C and T be as in Figure 34. Triangles ACQ and ATB are right triangles sharing

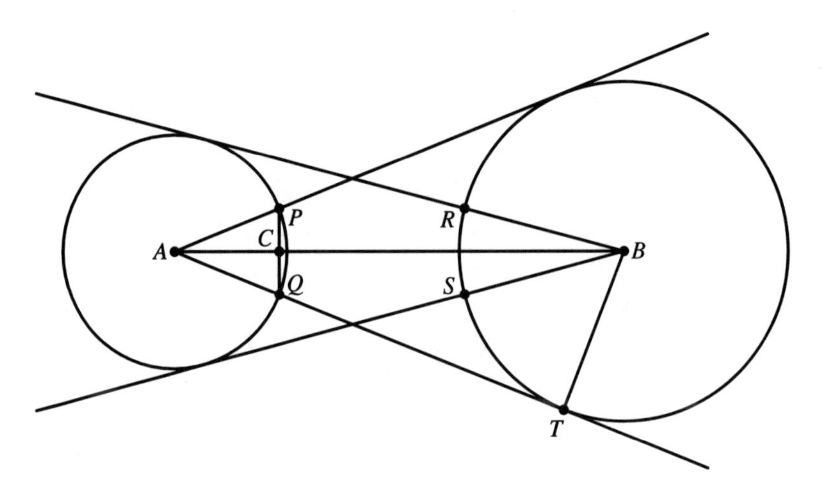

FIGURE 34

the angle BAT, so they are similar. Therefore

$$\frac{CQ}{r_A} = \frac{CQ}{AQ} = \frac{TB}{AB} = \frac{r_B}{AB},$$

so $CQ = r_A r_B / AB$, and therefore $PQ = 2r_A r_B / AB$.

A similar calculation leads to the same formula for RS, so $PQ = RS$.

34. Five Circles

Call the circles C_1, C_2, C_3, C_4, and C_5, reading from the left. Let a, b, c be the radii of C_2, C_3, C_4, respectively, let P be the point of intersection of the two tangent lines, and let p_1, p_2 be the distances from P to the centers of C_1, C_2, respectively.

Consider the transformation that shrinks the entire figure towards P by a factor of p_1/p_2 (that is, a rescaling using P as the origin). This transformation turns circles into (smaller) circles, preserves tangency, maps each of the two tangent lines into itself, and takes C_2 to where the original C_1 was. Moreover, since all the tangencies are preserved, each transformed C_i sits atop the old C_{i-1}. But this means that each of the ratios $c/9$, b/c, a/b, and $4/a$ equals p_1/p_2. Therefore $(c/9)(b/c) = (a/b)(4/a)$, which yields $b = 6$. (The sequence of five radii forms a geometric progression with ratio $\sqrt{3/2}$.)

35. A Ring of Disks

Let the centers of the disks be A, B, C, and D, and let the four points of tangency be P, Q, R, and S. Note that triangles APS, BQP, CRQ, and DSR are all isosceles. Therefore

$$\begin{aligned}\angle RSP + \angle RQP &= (180° - \angle DSR - \angle ASP) + (180° - \angle CQR - \angle BQP)\\ &= (180° - \angle DRS - \angle APS) + (180° - \angle CRQ - \angle BPQ)\\ &= (180° - \angle DRS - \angle CRQ) + (180° - \angle APS - \angle BPQ)\\ &= \angle SRQ + \angle SPQ.\end{aligned}$$

But then since the interior angles of a quadrilateral must sum to 360°, we must have $\angle RSP + \angle RQP = \angle SRQ + \angle SPQ = 180°$. The desired result now follows from the fact that if the sums of opposite angles in a quadrilateral are 180° then the quadrilateral can be inscribed in a circle.

Notes. It is possible to arrange four disks in a ring so that the quadrilateral formed by their centers is not convex. In this case the proof we have given will require some slight modifications, but the conclusion still holds.

An alternative solution begins with the fact that $AB + CD = AD + BC$, since both are equal to the sum of the radii of the four disks. It follows that a circle can be inscribed in quadrilateral $ABCD$.

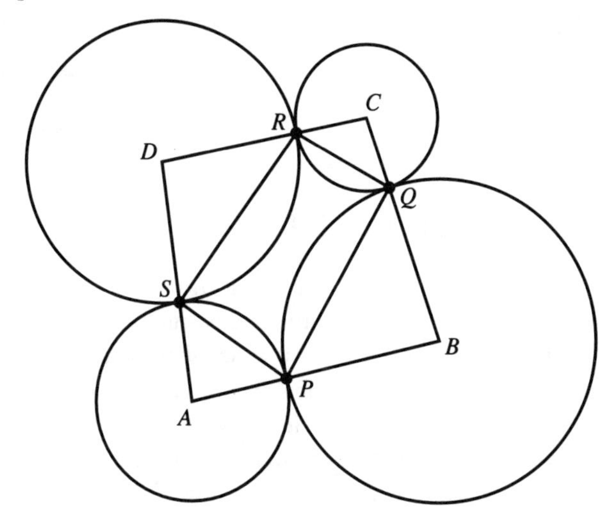

FIGURE 35

Problem 35.1. Show that the center of the circle inscribed in $ABCD$ is also the center of the circle that passes through P, Q, R, and S.

36. A Circle Inside an Angle

Draw a line through the center of the circle, parallel to the bisector of the angle. Then the points L and G where the sum of the distances to the sides of the angle are least and greatest are the points of intersection of the circle with this line. (Note that these are *not* necessarily the points that are closest to and furthest from the vertex of the angle.)

To prove that these points do the job, consider a line segment AB perpendicular to the bisector of the angle, and any point P on AB. Let the measure of the angle at the vertex O be α. Then $\angle OAB = \angle OBA = (\pi - \alpha)/2$, and it follows that the sum of the distances from P to the sides of the angle is $AP \sin\big((\pi - \alpha)/2\big) + PB \sin\big((\pi - \alpha)/2\big) = AB \sin\big((\pi - \alpha)/2\big)$. Therefore this sum is the same for any point P on AB, and it is proportional to AB, so it increases as A and B move away from O. It should now be clear that the desired points are the two points of tangency of lines perpendicular to the angle bisector with the circle, and these are the points L and G found earlier.

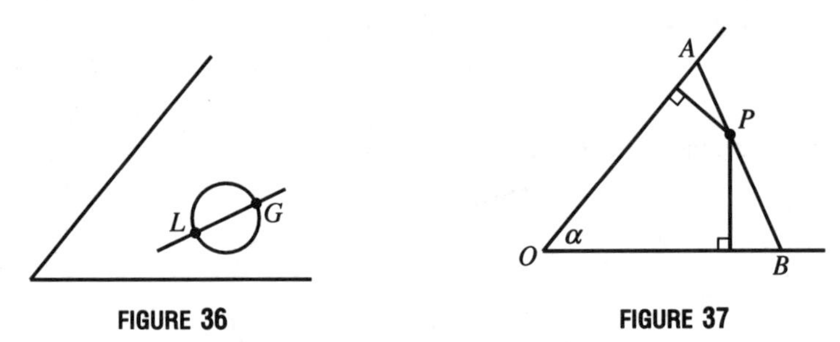

FIGURE 36 FIGURE 37

37. A Ray that Pierces Concentric Circles

It is useful to turn the problem on its head. Fix R on the outer circle, say at its very top, with P varying on a third concentric circle, smaller than the other two. Which choice of P maximizes QR? It is easy to see that the solutions to this auxiliary problem are the two points P such that RP is tangent to the small circle at P. Such a choice of P makes $\angle RPO$ a right angle, where O is the common center.

But this solves the original problem too. For suppose now that P is fixed at a point so that $\angle RPO$ is $90°$ when R is at the top of the circle. We claim that this maximizes QR. For suppose R' is another point on the outer circle for which $Q'R'$ is larger than QR. Let ρ be the rotation about the center that carries R' into

R and consider $\rho(P)$, $\rho(Q')$, and R: $\rho(Q')R = Q'R'$, which is longer than QR, contradicting the auxiliary result since P is the optimal point on the inner circle.

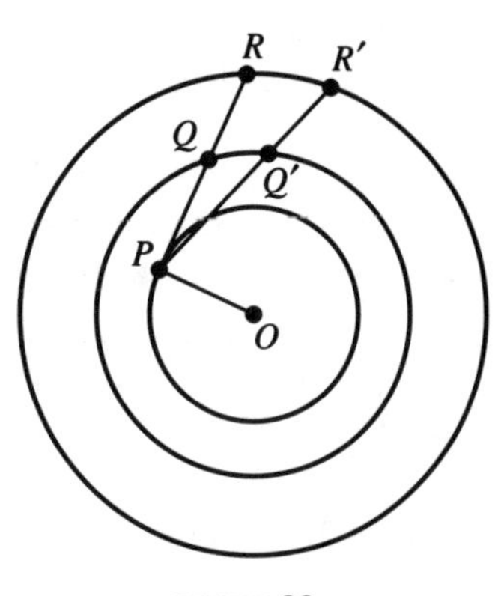

FIGURE 38

Alternate Solution. Let $RQPST$ be the chord of the larger circle with S on the smaller circle and T on the larger. An ancient proposition (Euclid, III.35) tells us that $QR \cdot QT$ is constant. Write QT as $QS + ST$ and observe that $QR = ST$, so that $QR \cdot QT = QR(QS + ST) = QR(QS + QR)$ is constant. To maximize QR we need to minimize QS, and QS is least when P is the midpoint of QS; this again follows from Euclid, III.35. Now, P bisects QS when and only when PR is perpendicular to OP.

38. If You Lose Your Compass

Let Q and R be the intersections of the circle with the lines AP and BP, respectively; let X be the intersection of AR with BQ. Because $\angle AQB = \angle ARB = 90°$, QX and PR are altitudes of $\triangle APX$. Because these altitudes pass through B, B must be the intersection point of all three altitudes, and therefore AB is an altitude of $\triangle APX$, which tells us that PX is the desired perpendicular. If P lies over or within the circle then the construction is the same but the proof is a

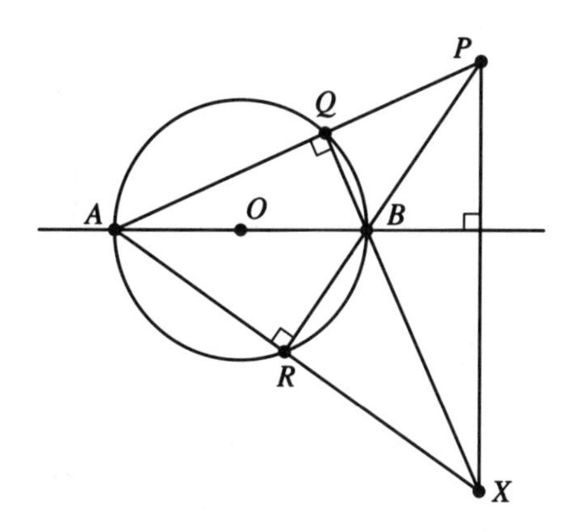

FIGURE 39

little different: X becomes the orthocenter of $\triangle APB$ in the first case, while if P is inside the circle, P is the orthocenter of $\triangle ABX$.

Note. This problem was given in the Old Farmer's Almanac [Sch]. If P is on the circle then the preceding method does not work. But a famous theorem of Steiner (1833; see [Dor]) states that any straight-edge and compass construction can actually be carried out with straight-edge alone, provided the plane contains a single circle and its center. Thus there is in fact a method to solve the case that P is on the circle too.

Problem 38.1. Solve Problem 38 when P is on the circle.

39. Circles in a Circle

We use a coordinate system in which the large circle is centered at the origin and has radius 1, and AB lies along the x-axis. Then CD lies along the line $x = d$, for some d between -1 and 1.

First consider circle 3. Letting O be the origin, P the center of circle 3, and r its radius, it is clear from Figure 40 that $P = (d+r, -r)$ and the distance from O to P is $1-r$. Thus $(d+r)^2+r^2 = (1-r)^2$, so $r^2+(2+2d)r+d^2-1 = 0$, and $r = -1-d\pm\sqrt{2+2d}$; since r must be positive it follows that $r = -1-d+\sqrt{2+2d}$. Similar reasoning shows that the radius of circle 2 is $-1+d+\sqrt{2-2d}$.

To find the radius of circle 1, we use the following lemma.

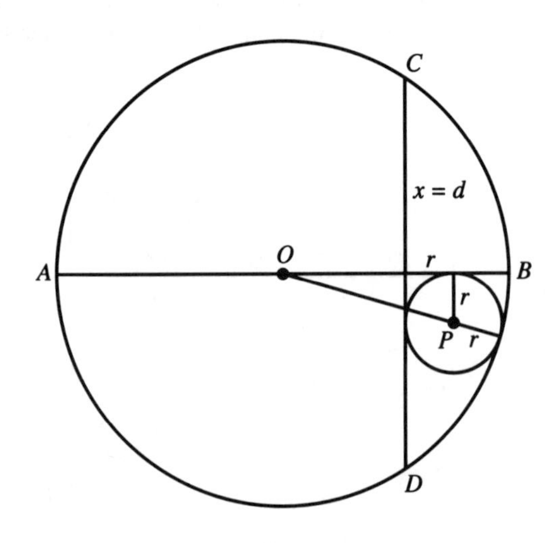

FIGURE 40

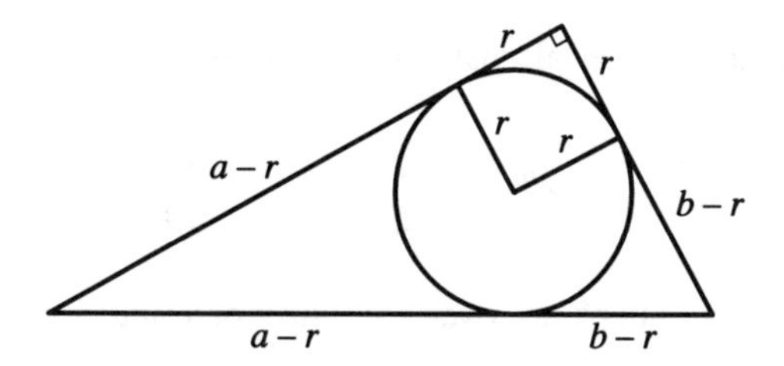

FIGURE 41

Lemma. *If a circle is inscribed in a right triangle whose legs have lengths a and b and whose hypotenuse has length c, then the radius of the inscribed circle is $(a+b-c)/2$.*

Proof. Let the radius of the inscribed circle be r. Then it is clear from Figure 41 that $c=(a-r)+(b-r)$, and the desired conclusion follows.

Since the coordinates of A, B, and C, respectively, are $(-1,0)$, $(1,0)$, and $(d,\sqrt{1-d^2})$, the legs of $\triangle ABC$ have lengths $\sqrt{2+2d}$ and $\sqrt{2-2d}$ and the hypotenuse has length 2. Therefore by the lemma the radius of circle 1 is $(\sqrt{2+2d}+\sqrt{2-2d}-2)/2$, which is the average of the radii of circles 2 and 3.

Problem 39.1. Show that if lines are drawn through the centers of circles 2 and 3 parallel to CD, then these lines are tangent to circle 1. This gives an alternative solution to the problem, since the distance between these lines is then equal to both the diameter of circle 1 and also the sum of the radii of circles 2 and 3.

Note. For related problems, see [FP, example 2.4(2)], [Lar, problem 8.1.15], and [Kat].

40. Get Close to the Circles

Let P be the origin of a rectangular coordinate system, and let the coordinates of O_i in this coordinate system be (x_i, y_i). Then

$$(PO_1)^2+(PO_2)^2+(PO_3)^2+(PO_4)^2=\sum_i (x_i^2+y_i^2)=\sum_i x_i^2+\sum_i y_i^2,$$

where in the summations i ranges from 1 to 4.

Since the circles are nonoverlapping and have unit radius, for each $i \neq j$ we have

$$(O_iO_j)^2=(x_i-x_j)^2+(y_i-y_j)^2 \geq 4.$$

Expanding and summing over all distinct i and j yields

$$3\sum_i x_i^2 + 3\sum_i y_i^2 - 2\sum_{i,j} x_i x_j - 2\sum_{i,j} y_i y_j \geq 24,$$

where the double summations are over all values of i and j with $1 \leq i < j \leq 4$. Adding

$$\sum_i x_i^2 + \sum_i y_i^2 + 2\sum_{i,j} x_i x_j + 2\sum_{i,j} y_i y_j$$

to both sides we get

$$\begin{aligned} 4\sum_i x_i^2 + 4\sum_i y_i^2 &\geq \sum_i x_i^2 + \sum_i y_i^2 + 2\sum_{i,j} x_i x_j + 2\sum_{i,j} y_i y_j + 24 \\ &= \left(\sum_i x_i\right)^2 + \left(\sum_i y_i\right)^2 + 24 \geq 24. \end{aligned}$$

Thus

$$(PO_1)^2 + (PO_2)^2 + (PO_3)^2 + (PO_4)^2 = \sum_i x_i^2 + \sum_i y_i^2 \geq 6,$$

as required.

Problem 40.1. Can you improve on the lower bound of 6?

41. An Ellipse and a Circle

If we take the major and minor axes of the ellipse to lie along the x- and y-axes, then the ellipse is given by the equation

$$\frac{x^2}{a^2} + \frac{y^2}{b^2} = 1.$$

Note that if the tangent lines are horizontal or vertical then the conclusion is obvious, so we assume without loss of generality that, as in Figure 42, the points of tangency of the lines and the ellipse are in the interiors of the second and fourth quadrants. Let the coordinates of the point of tangency in the second quadrant be (x_0, y_0).

To find the radius of the circle, we project the vector $x_0\mathbf{i} + y_0\mathbf{j}$ onto a unit vector $\mathbf{u}$ perpendicular to the tangent lines. If the angle of inclination of the tangent lines

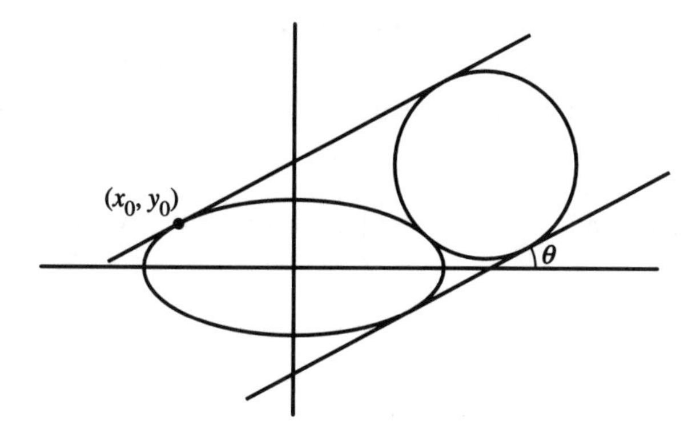

FIGURE 42

is θ then we can take $\mathbf{u} = -(\sin\theta)\mathbf{i} + (\cos\theta)\mathbf{j}$, so the radius r of the circle is given by $r = (x_0\mathbf{i} + y_0\mathbf{j}) \cdot \mathbf{u} = -x_0 \sin\theta + y_0 \cos\theta$.

Note that by implicit differentiation the slope of the tangent line to the ellipse at any point (x, y) on the ellipse is given by

$$\frac{dy}{dx} = -\frac{b^2 x}{a^2 y},$$

so we have $\tan\theta = -b^2 x_0/(a^2 y_0)$, or in other words

$$a^2 y_0 \sin\theta = -b^2 x_0 \cos\theta.$$

Applying this and the fact that (x_0, y_0) is on the ellipse, we find that

$$\begin{aligned} r^2 &= x_0^2 \sin^2\theta - 2x_0 y_0 \sin\theta\cos\theta + y_0^2 \cos^2\theta \\ &= (x_0^2 \sin^2\theta - x_0 y_0 \sin\theta\cos\theta) + (y_0^2\cos^2\theta - x_0 y_0 \sin\theta\cos\theta) \\ &= (x_0^2 \sin^2\theta + (a^2/b^2) y_0^2 \sin^2\theta) + (y_0^2 \cos^2\theta + (b^2/a^2) x_0^2 \cos^2\theta) \\ &= a^2 \sin^2\theta (x_0^2/a^2 + y_0^2/b^2) + b^2 \cos^2\theta (x_0^2/a^2 + y_0^2/b^2) \\ &= a^2 \sin^2\theta + b^2 \cos^2\theta, \end{aligned}$$

so we have another formula for the radius of the circle: $r = \sqrt{a^2 \sin^2\theta + b^2 cos^2\theta}$.

Clearly if the distance between the centers of the ellipse and circle is to be $a + b$, then the center of the circle must be the point $\big((a+b)\cos\theta, (a+b)\sin\theta\big)$, which we denote by P. So what we must show is that the circle centered at P with radius r is tangent to the ellipse. But using our second formula for r, it is easy to see that the point $(a\cos\theta, b\sin\theta)$ is on both this circle and the ellipse, so it will suffice to check that the slopes of the tangent lines to the circle and ellipse at this point are the same. Elementary calculus shows that both slopes are $-b\cos\theta/(a\sin\theta)$, so we are done.

Note. This problem is a Japanese temple problem [FP, example 6.1].

42. Three Circles in a Circle

Let the small circles have radius 1, and let the radius of the big circle be R. Let X, Y, Z, and O be the centers of the three small circles and the big circle. Then $\triangle XYZ$ is an equilateral triangle with side-length 2, and O is in the center of this triangle. It follows that $OX = OY = OZ = 2\sqrt{3}/3$, so R is $1 + 2\sqrt{3}/3$, or $(3+2\sqrt{3})/3$.

We first compute PA. Since $\angle PAX = 90°$ and $AX = 1$, we have $PA^2 = PX^2 - 1$. To compute PX^2, we use the law of cosines. Let $\angle POX$ be θ. Then $PX^2 = PO^2 + OX^2 - 2 \cdot PO \cdot OX \cdot \cos\theta$. But PO is $(3+2\sqrt{3})/3$ and $OX = 2\sqrt{3}/3$. Substitution yields $PX^2 = (11+4\sqrt{3})/3 - (8+4\sqrt{3})/3\cos\theta$. So

$$PA^2 = PX^2 - 1 = \frac{11+4\sqrt{3}}{3} - \frac{8+4\sqrt{3}}{3}\cos\theta - 1$$
$$= \frac{8+4\sqrt{3}}{3} - \frac{8+4\sqrt{3}}{3}\cos\theta$$
$$= \frac{8+4\sqrt{3}}{3}(1-\cos\theta) = \frac{8+4\sqrt{3}}{3}(2\sin^2(\theta/2)),$$

and finally

$$PA = \sqrt{\frac{16+8\sqrt{3}}{3}}\sin(\theta/2).$$

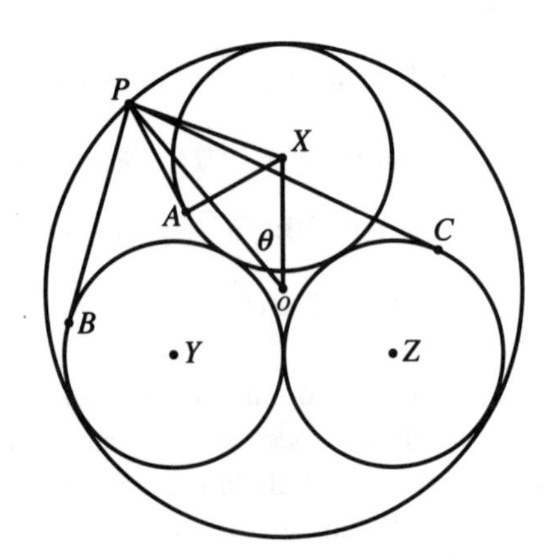

FIGURE 43

We can now let $d = \sqrt{(16+8\sqrt{3})/3}$; then $PA = d\sin(\theta/2)$. Similarly we get $PB = d\sin(\rho/2)$, and $PC = d\sin(\phi/2)$, where ρ and ϕ are the obvious angles. So we must confirm that $\sin(\theta/2) + \sin(\rho/2) = \sin(\phi/2)$. But it is clear that $\rho = 2\pi/3 - \theta$ and $\phi = 2\pi/3 + \theta$, so it suffices to shows that

$$\sin\frac{\theta}{2} + \sin\left(\frac{\pi}{3} - \frac{\theta}{2}\right) = \sin\left(\frac{\pi}{3} + \frac{\theta}{2}\right).$$

But this is straightforward:

$$\begin{aligned}\sin\frac{\theta}{2} + \sin\left(\frac{\pi}{3} - \frac{\theta}{2}\right) &= \sin\frac{\theta}{2} + \sin\frac{\pi}{3}\cos\frac{\theta}{2} - \cos\frac{\pi}{3}\sin\frac{\theta}{2} \\ &= \sin\frac{\theta}{2} + \frac{\sqrt{3}}{2}\cos\frac{\theta}{2} - \frac{1}{2}\sin\frac{\theta}{2} \\ &= \frac{1}{2}\sin\frac{\theta}{2} + \frac{\sqrt{3}}{2}\cos\frac{\theta}{2} \\ &= \sin\left(\frac{\pi}{3} + \frac{\theta}{2}\right).\end{aligned}$$

Note. This result is a special case of Casey's Theorem, which as generalized by Wigand, states: *Given a regular n-gon, with n odd and vertices $v_1, \ldots, v_n$, and C its circumcircle. At each v_i draw a circle that is internally tangent to C at v_i, and suppose all these tangent circles are congruent. Let P be any point on the minor arc from v_1 to v_n and let t_i be the length of the tangent from P to the circle tangent to C at v_i. Then $\sum_{i=1}^{n}(-1)^i t_i = 0$.* For proofs and more information, see [Rab5].

43. Two Nickels and Three Pennies

For pennies and nickels the difference in the two arrangements is very, very small, about one ten-thousandth of an inch. Nevertheless, it turns out that for any (unequal) sizes of the disks representing pennies and nickels, the arrangement with adjacent nickels leads to a larger interior disk. Consider Figure 44, which presents a wild exaggeration of the relative sizes of the coins, and look

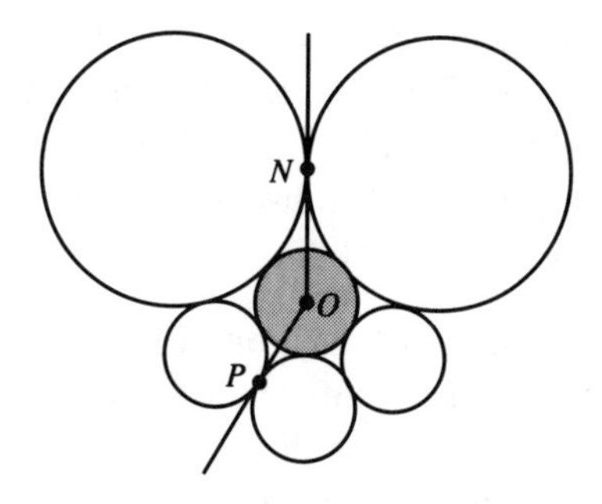

FIGURE 44

at the wedge formed by the lines from O, the center of the interior disk, to N (the nickel/nickel tangent point) and P (the penny/penny tangent point indicated). If this wedge is flipped about the line bisecting the angle at O, Figure 45 results. Such a reflection interchanges the lines, and a close look shows that the new positions of N and P are not tangent to their new neighbors. This follows from $ON \neq OP$, which we will prove in a moment. This will solve the general problem, for it is clear that the interior disk will have to be shrunk to restore the tangency.

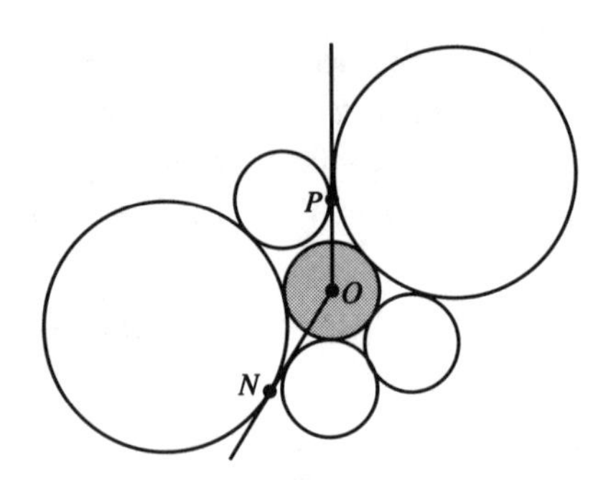

FIGURE 45

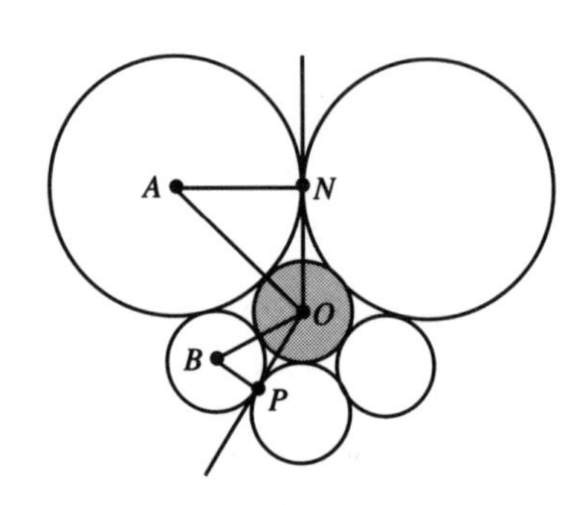

FIGURE 46

It remains to show that $ON \neq OP$. But this follows immediately from the fact that $\triangle ONA$ and $\triangle OPB$ have right angles at N and P, respectively. Now, if ON^2 did equal OP^2, and if p, n, and r denote the radii of the pennies, nickels, and interior disk, respectively, then the theorem of Pythagoras would yield that $(r+n)^2-n^2=(r+p)^2-p^2$, or $p=n$, contradiction.

Problem 43.1. The official radius of a U.S. penny is 9.525 mm; for a nickel it is 10.605 mm. Show that, using these dimensions, the difference in the radii of the two interior tangent circles that arise from the two types of rings is approximately 1/400 of a millimeter (a mere 1/10,000 of an inch).

Problem 43.2. Garry Ford has observed [Gar2, p. 164] that six dimes and a quarter fit almost exactly around a quarter (see color plate 2 following page 16). A dime's radius is 8.955 mm and a quarter's is 12.13 mm. Is the interior disk larger or smaller than a quarter? (In fact, because of manufacturing tolerances it is possible to find dimes and quarters for which the fit is perfect!)

Notes. There is a bit of hand-waving going on here, since we never proved that the tangent configuration exists, nor that it is unique. Here is a proof. Let c, n, and p denote the radii of the interior disk we seek, the penny, and the nickel, respectively. Look at a single tangent configuration made up of the central disk, a penny, and a nickel. The three centers form a triangle with sides $c+n$, $c+p$, and

$p + n$; let θ be the angle at the center of the disk with radius c. Applying the law of cosines to this angle and simplifying gives

$$\cos\theta = 1 - \frac{2np}{c^2 + cn + cp + np}$$

which is monotonic in c. Moreover, as c approaches infinity, θ approaches 0 and as c approaches 0, θ approaches π.

Now, we are trying to fit 5 such triangles (penny/nickel, nickel/nickel, or penny/penny triangles) together so that the angles called θ sum to 2π. But $\sum_{i=1}^{5} \theta_i(c)$ is a monotonic function of c that lies between 0 and 5π as c shrinks from infinity to 0. Therefore there is a unique value of c for which this sum is 2π.

This argument works for n coins, if $n > 2$. For 2 coins, an interior tangent disk does not exist, and the proof shows why: two angles each less than π cannot sum to 2π. There is a slight flaw here, however, for suppose the coin radii are 100, 1, 100, 100, 100, and 100. Then once the teeny coin is placed beside a large one, another 100-radius coin cannot go beside the teeny one without crashing into the first 100-radius coin. It is not clear that there is a nice condition that eliminates this problem in general, so it is perhaps simplest, when considering the n-coin problem, to allow overlap of coins. Now, one can ask, given n coins of varying, not necessarily distinct, sizes, for the configuration that maximizes (or minimizes) the radius of the interior disk. The answer [DVW] is that if the n radii are $r_1, r_2, \ldots, r_n$, with $r_1 \geq r_2 \geq r_3 \geq \cdots \geq r_n$, then the configuration that maximizes the interior radius is $\ldots, r_5, r_3, r_1, r_2, r_4, \ldots$. To minimize, use $\ldots, r_5, r_{n-3}, r_3, r_{n-1}, r_1, r_n, r_2, r_{n-2}, r_4, r_{n-4}, \ldots$.

44. Thirteen Bottles of Wine

Label the centers A, B, C, and so on, and draw the connecting lines illustrated in Figure 47. Since D is at distance 2 from A, B, and F, the radius-2 circle centered at D passes through A, B, and F. And since $\angle BAF = 90°$, BF is a diameter and so BDF is a straight line. Now, because $BDGE$ and $EGJH$ are rhombi, $HJ||BD$; similarly $JL||DF$. So HJL is a straight line. Draw the radius-2 circle centered at J; it contains L, H, and M whence $\angle LMH$ is 90° and, because HM is vertical, LM is horizontal. Similarly KL is horizontal.

Notes. The solution does not explicitly use the disjointness of the circles. In fact, if $\sqrt{12} < AB < 4$, then circle f will have to overlap circle a if the condition of being tangent to d and the y-axis is to hold. But the proof still works. The wine-bottle problem is due to Charles Payan of the Laboratoire de Structures Dèscretes et de Didactique in France, one of the creators of the remarkable package for plane

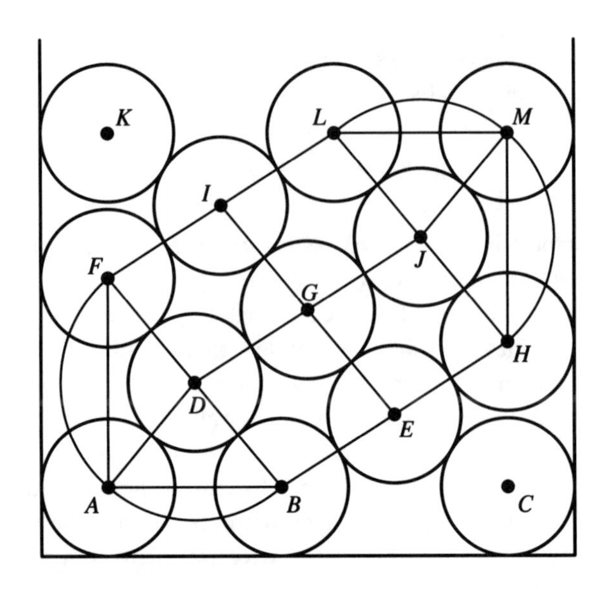

FIGURE 47

geometry called *CABRI.* Payan discovered this result while using the software. The proof given here is due to Hung Dinh.

Problem 44.1. Find and prove a version of Problem 44 for a bottom row having n circles.

45. A Hexagon in a Circle

Let O be the center of the circle and let angles and distances be as in Figure 48. Because $\triangle OPQ$ is isosceles, $\theta = (180° - \alpha)/2$. Similarly, $\rho = (180° - \beta)/2$. Therefore $\theta + \rho = 180° - (\alpha + \beta)/2 = 180° - 60° = 120°$. This means that $\theta + \rho = \alpha + \beta$. Now apply the law of cosines to $\triangle POR$ to get $d^2 = r^2 + r^2 - 2r^2 \cos(\alpha + \beta) = 3r^2$. And do the same to $\triangle PQR$ to get $d^2 = 1^2 + a^2 - 2a \cos(\theta + \rho) = 1 + a + a^2$. Combining these equations tells us that $r = \sqrt{(1 + a + a^2)/3}$.

For $1 + a + a^2$ to be divisible by 3, a must have the form $3k + 1$. And for the quotient to be a perfect square, $3k^2 + 3k + 1$ must be a square. At this point, trial and error will lead to the smallest solution (excluding $a = 1$), which comes from $k = 7$ and yields $a = 22$ and $r = 13$. To solve the general problem, suppose $3k^2 + 3k + 1 = m^2$. We complete the square to get $3(2k + 1)^2 = (2m)^2 - 1$ or $(2m)^2 - 3(2k + 1)^2 = 1$. This has the form $x^2 - 3y^2 = 1$, which is a type of

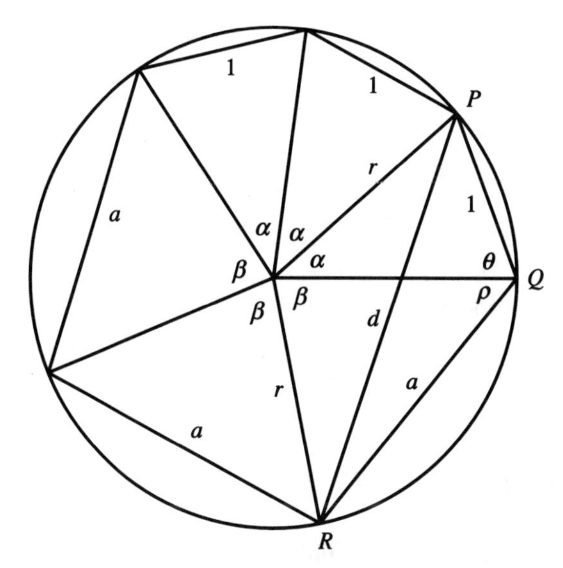

FIGURE 48

equation known as a Pell equation. Standard techniques ([NZM, theorem 7.26]) tell us that the y values of solutions are given by the coefficients of $\sqrt{3}$ in integer powers of $2+\sqrt{3}$. And the first three solutions are $(x,y)=(2,1)$, $(x,y)=(7,4)$, and $(x,y)=(26,15)$. This last corresponds to $k=7$, $r=13$, a solution already mentioned.

If we now let $(2+\sqrt{3})^n = x_n + y_n\sqrt{3}$, it is easy to check that $x_n = 2x_{n-1}+3y_{n-1}$ and $y_n = x_{n-1}+2y_{n-1}$. This reduces to $y_n = 4y_{n-1} - y_{n-2}$, which tells us, because $y_1 = 1$ and $y_2 = 4$, that the y-sequence has alternating parity. Since we are interested in odd values of y_n (from $y = 2k+1$ above), we are interested in y_n where n is odd (and note that if $x^2-3y^2=1$ and y is odd, then x is necessarily even). This proves that there are infinitely many values of a for which the circle's radius is an integer. A short *Mathematica* computation yields the following table.

```
y[n_] := (y[n] = 4 y[n - 1] - y[n - 2])
y[1] = 1;
y[2] = 4;
k[n_] := (y[n] - 1)/2
a[n_] := 3 k[n] + 1
r[n_] := Sqrt[(1 + a[n] + a[n]^2) / 3]
TableForm[Table[{y[n], k[n], a[n], r[n]}, {n, 1, 13, 2}]]
```

y	k	a	r
1	0	1	1
15	7	22	13
209	104	313	181
2911	1455	4366	252
40545	20272	60817	35113
564719	282359	847078	489061
7865521	3932760	11798281	6811741

7.5 Packing and Covering

46. A 6-Sided Peg in a Square Hole

The largest such hexagon is one having a diagonal that lies along a diagonal of the square. If the square is $ABCD$ and has center O, draw EH and GJ so that they pass through O and make angles of 60° with AC, as illustrated. Choosing F and I in the proper position along AC yields the hexagon $EFGHIJ$. To see that this is the largest regular hexagon that fits, observe that if it is rotated less than 30° to the left then the diagonal EH will decrease, while if it is rotated less than 30° to the right the diagonal GJ will decrease.

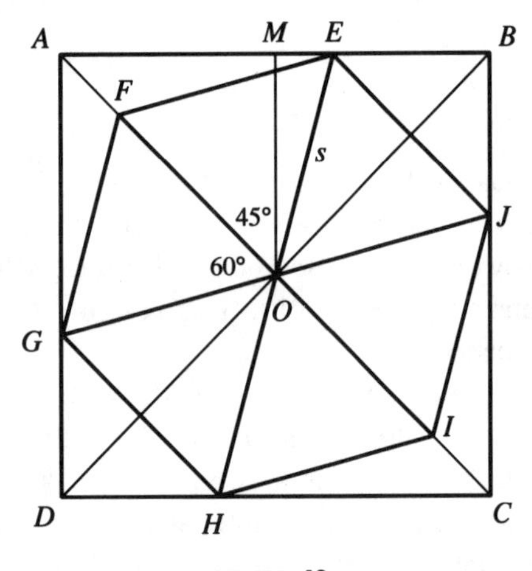

FIGURE 49

To get s, the side-length of the hexagon, draw the perpendicular from O to the top of the square at M and observe that $\angle MOE = 15°$. Therefore

$$OE = \sec 15° = \sqrt{6} - \sqrt{2}.$$

The hexagon's area is then $12\sqrt{3} - 18$, which is just under 70% of the area of the square.

Note. This problem is due to Cleon Richtmeyer.

47. Packing 11 Squares

This problem is unsolved. It is natural to first consider 45° packings, by which we mean configurations in which each unit square is either aligned with or tilted at 45° to the large square. The best known 45° packing is due to F. Göbel [Gob] and is given in the diagram on the left, which packs 11 unit squares into a square whose diagonal has length $2 + 2\frac{1}{2}\sqrt{2}$. The side of the large square therefore has length $\frac{5}{2} + \sqrt{2}$ (≈ 3.914). However, this configuration is not the overall champion. In 1979 Walter Trump (Germany) found a packing into a square of side-length 3.877... (see the diagram on the right).

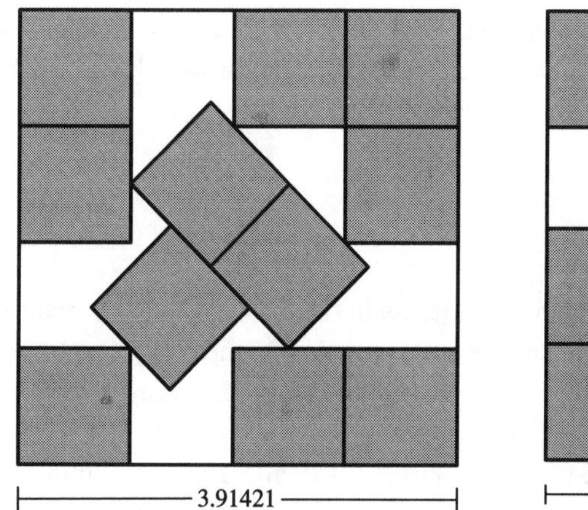

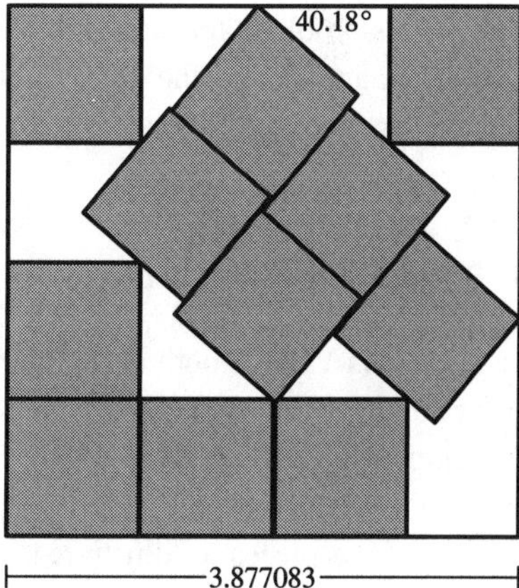

FIGURE 50

In 1984 Walter Stromquist (Daniel H. Wagner Associates, Paoli, PA) proved (unpublished) that Trump's configuration cannot be improved by a 45° packing. He also showed that the packings of 1–10 squares given by Martin Gardner in [Gar2, chapter 20] are optimal; these packings are all 45° packings. This means that the case of 11 squares is the first where a non-45° packing is champion (even though there is as yet no proof of optimality of the Trump example). See [Gar2, chapter 20] for a discussion of many results and questions in this area.

Problem 47.1. Verify the two constants (3.87... and 40.18...) in Trump's non-45°-packing. Hint: Call the constants k and θ and find two equations relating them; the equations reduce to a single equation in θ that can be solved by standard rootfinding.

48. Find the Rectangles

Assume the 400-gon has a horizontal bottom edge. Look at a path of tiles from the top edge to the bottom edge. That path consists of tiles with horizontal tops and bottoms. Then look at a path from the leftmost edge to the right most edge; its tiles have vertical sides. But these two paths must cross; and the tile at which they meet is therefore a rectangle. This argument can be carried out 100 times, yielding a new rectangle each time.

Problem 48.1. Show that it is indeed possible to tile a regular 400-gon with parallelograms.

Note. This problem appeared in the spring, 1983, Tournament of the Towns [Tay, pp. 67–69], and also in the 1987 Lower Michigan Math Contest, which was set by Stephen Landsberg (University of Rochester).

49. Straightening Out a Circle

The answer is YES to both parts, with solutions as in the diagrams. A key point is the fact that for any two closed line segments or closed arcs, there is a one-to-one continuous mapping of the points of one onto the points of the other. Now, for the triangle we divide one side into two pieces and draw lines from another side into one of the pieces as illustrated. For the circle we divide a diameter into five pieces and use two of them to generate segments that cover the two semicircles, as illustrated.

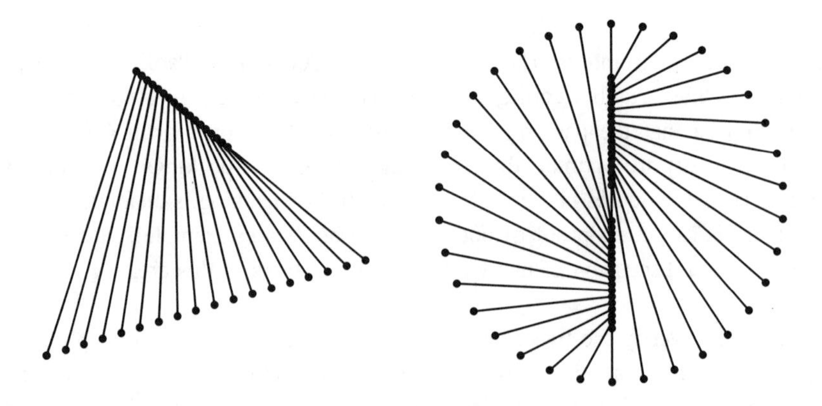

FIGURE 51

7.6 Area

50. A Japanese Temple Problem

Add lines as in Figure 52, and let the lengths of the sides of the two untilted squares be a and b. Note that triangles ABE, BCH, EFK, and GHK are all congruent, so $EF = GK = BC = AE = a$ and $FK = GH = AB = CH = b$. Also, triangles DFK and JKM are congruent, as are GIK and KLN, so $JM = DF = 2a$, $JK = FK = b$, $KL = GK = a$, and $LN = GI = 2b$.

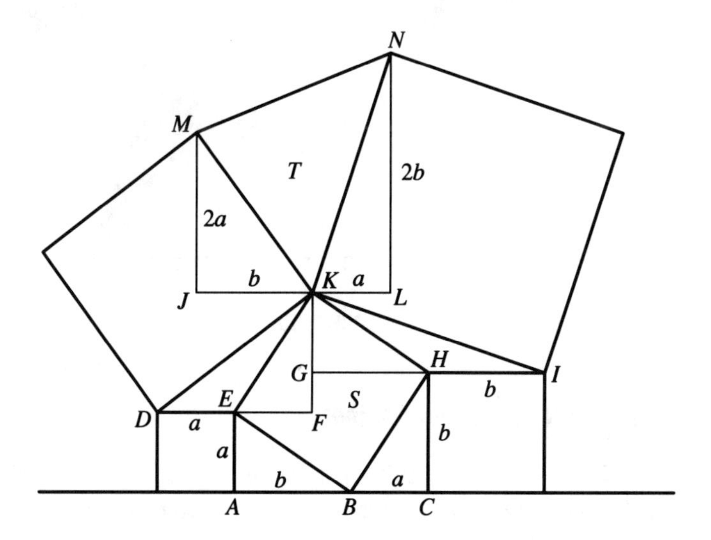

FIGURE 52

We can now compute the areas of S and T:

$$\begin{aligned}\text{area}(T) &= \text{area}(JLNM) - \text{area}(JKM) - \text{area}(KLN)\\ &= (a+b)(2a+2b)/2 - ab - ab\\ &= a^2 + b^2 = EB^2\\ &= \text{area}(S).\end{aligned}$$

Note. This problem was written in 1844 on a tablet in the Aichi prefecture [FP, problem 4.2.4, pp. 48, 132].

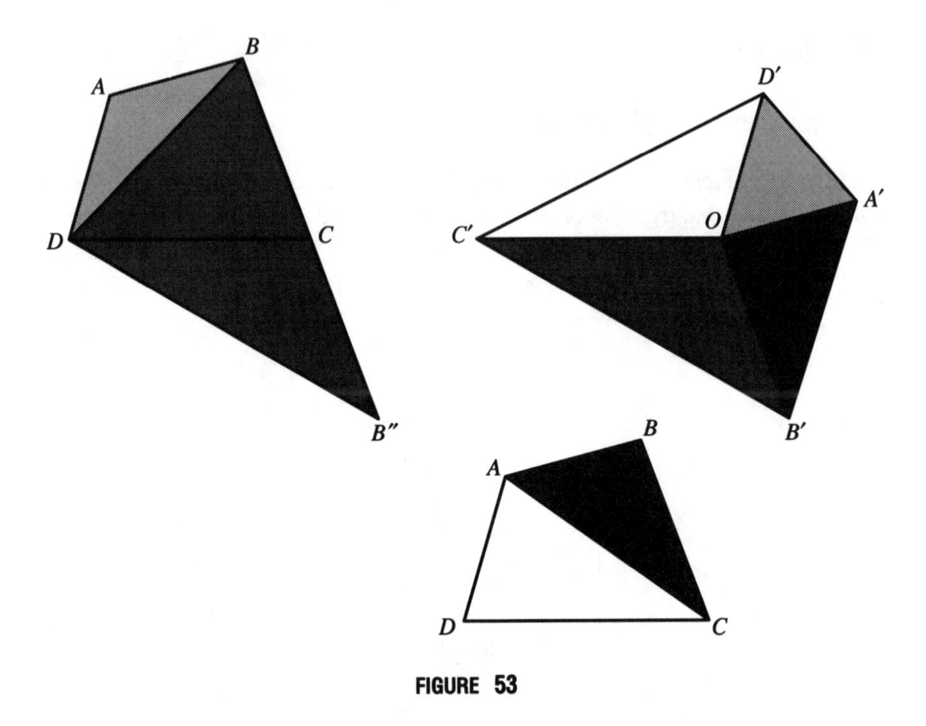

FIGURE 53

51. One Quadrilateral Begets Another

Since $DC = OC'$, $BC = OB'$, DC is parallel to OC', and BC is parallel to OB', triangles BCD and $B'OC'$ have the same area (they can be viewed as having equal bases (BC, OB') and altitudes (from D or C'; see $\triangle B''CD$ in Figure 53). The other pairs of triangles indicated in the diagram have equal area as well, and so quadrilateral $A'B'C'D'$ has twice the area of $ABCD$.

52. Rectangles in a Rectangle

Let O be the midpoint of the segment HF. Clearly O is the center of both $EFGH$ and $IFJH$, so $OE = OH = OJ$. Triangles AEH and CGF are congruent, so $AH = CF$, and therefore O is the center of $ABCD$. But then since $OE = OJ$, it follows that $AE = DJ$, so EJ is parallel to AD. Add the segments PF and HQ, parallel to AB.

Now consider the shaded quadrilateral $EFJH$. On the one hand it contains half of each of the rectangles $AEQH$, $EBFP$, $HQJD$, and $PFCJ$, so its area is one-half the area of $ABCD$. On the other hand, dividing it along the line HF we see that it consists of half of rectangle $EFGH$ and half of rectangle $IFJH$, so its

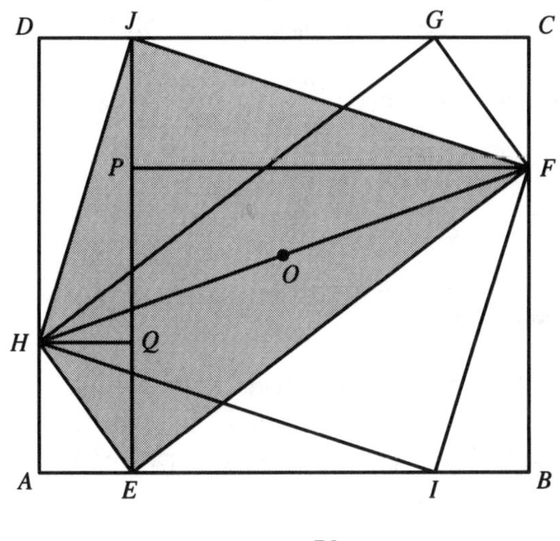

FIGURE 54

area is one-half the sum of the areas of these two rectangles. Therefore the sum of the areas of $EFGH$ and $IFJH$ is equal to the area of $ABCD$.

53. A Constant Difference

Let a denote the total area of $ABCD$. Label the 11 regions as in the diagram, and let a_i denote the area of region i. Since the area of $\triangle ABQ$ is $a/2$, we have $a_3 + a_4 + a_5 + a_1 + a_{10} + a_9 = a/2$, so $a_1 + a_3 + a_5 + a_9 = a/2 - a_4 - a_{10}$. Also, the areas of $\triangle ABP$ and $\triangle RBP$ are equal, so $a_7 = a_4 + a_{10} + a_{11}$. This tells us

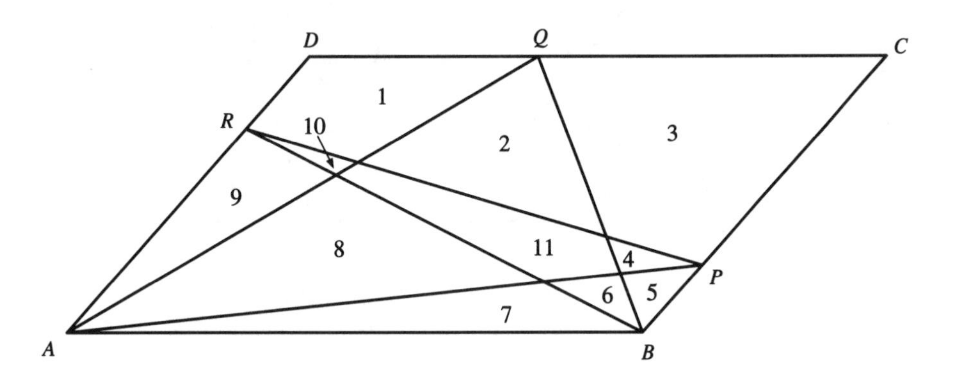

FIGURE 55

that the area of the red region minus the area of the blue region is

$$a_1 + a_3 + a_5 + a_7 + a_9 - a_{11} = \frac{a}{2} - a_4 - a_{10} + (a_4 + a_{10} + a_{11}) - a_{11}$$
$$= \frac{a}{2}.$$

54. An Area Comparison

The area of $\triangle ABC$ is always at least as large as the area of $\triangle DEF$. To see why, let d, e, and f be the distances from P to BC, AC, and AB respectively, let s be the length of a side of $\triangle ABC$, and let h be its altitude. Then

$$\frac{1}{2}sh = \text{area}(\triangle ABC) = \text{area}(\triangle BCP) + \text{area}(\triangle ACP) + \text{area}(\triangle ABP)$$
$$= \frac{1}{2}sd + \frac{1}{2}se + \frac{1}{2}sf$$
$$= \frac{1}{2}s(d + e + f).$$

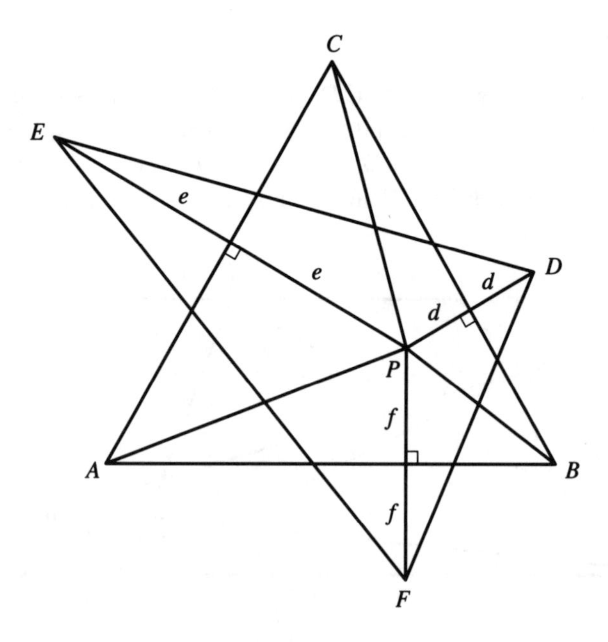

FIGURE 56

Thus $d+e+f=h=s\sqrt{3}/2$, so

$$\text{area}(\triangle ABC)=\frac{1}{2}sh=h^2\sqrt{3}/3=(d+e+f)^2\sqrt{3}/3.$$

To compute the area of $\triangle DEF$, first note that the angles of $\triangle ABC$ are all $60°$ and PD, PE, and PF are perpendicular to BC, AC, and AB, respectively; therefore $\angle DPE = \angle EPF = \angle DPF = 120°$. Thus the area of $\triangle DPE$ is $\frac{1}{2}(2d)(2e)\sin 120° = de\sqrt{3}$. Similarly, the areas of $\triangle EPF$ and $\triangle DPF$ are $ef\sqrt{3}$ and $df\sqrt{3}$, so $\text{area}(\triangle DEF)=(de+ef+df)\sqrt{3}$. Thus

$$\begin{aligned}\text{area}(\triangle ABC)-\text{area}(\triangle DEF)&=\left[(d+e+f)^2-3(de+ef+df)\right]\sqrt{3}/3\\&=\left[(d-e)^2+(e-f)^2+(d-f)^2\right]\sqrt{3}/6\geq 0,\end{aligned}$$

and $\text{area}(\triangle ABC)\geq\text{area}(\triangle DEF)$. In fact, it is clear from this solution that the areas are equal only if $d=e=f$, which happens only if P is the centroid of $\triangle ABC$.

55. Another Area Comparison

The areas are equal. To prove it, let a and b be the lengths of AC and BC, respectively. Then since triangles ACI and ADE are similar, $IC/a=b/(a+b)$, and therefore $IC=ab/(a+b)$. Similar reasoning gives the same value for JC. Thus,

$$\text{area}(\triangle ACI)=\frac{1}{2}a\frac{ab}{a+b}=\frac{a^2b}{2(a+b)}$$

and

$$\text{area}(\triangle BCJ)=\frac{1}{2}b\frac{ab}{a+b}=\frac{ab^2}{2(a+b)}.$$

Notice that the sum of these areas is

$$\frac{a^2b}{2(a+b)}+\frac{ab^2}{2(a+b)}=\frac{ab(a+b)}{2(a+b)}=\frac{ab}{2},$$

which is the same as the area of $\triangle ABC$! Since triangles ACI and BCJ cover all of ABC except for ABH, and they cover $HICJ$ twice, it follows that the areas of ABH and $HICJ$ are equal.

In the nonuniform case, we first use the fact that an inscribed circle in an equilateral triangle of side $2s$ has radius $s/\sqrt{3}$. Thus the large radius is $1/(2\sqrt{3})$. Now, the smaller circle is inscribed in an equilateral triangle of whose altitude is $1/(2\sqrt{3})$. Therefore the triangle's side is $1/3$ and its inscribed circle has radius $1/(6\sqrt{3})$. The total area is therefore $\pi(1/12 + 2/108)$.

Notes. The problem of inscribing three circles into an arbitrary triangle so as to maximize the total circular area is known as the Malfatti problem, after Giovanni Francesco Malfatti (1731–1807). Malfatti thought that the solution would always be when each circle was tangent to two other circles and to two sides, but, as this problem shows, this is not true for an equilateral triangle (this was not observed until 1930). In fact, as shown by M. Goldberg [Gol], in *every* triangle the proposed Malfatti solution fails to be optimal!

The Malfatti problem has been completely solved only recently [Gol1, ZL]. Here is how to optimize the area. Suppose the vertices of the given triangle are labelled A, B, C, with the corresponding angles α, β, and γ satisfying $\alpha \leq \beta \leq \gamma$. Now let the first circle, K_1, be the familiar inscribed circle in the triangle. Let K_2 be the circle tangent to AB, AC, and K_1. If $\sin(\alpha/2) \geq \tan(\beta/4)$, then K_3 is taken tangent to AB, BC, and K_1; otherwise K_3 touches AB, AC, and K_2. These three circles are the optimal configuration. It is noteworthy that the method described is, in fact, just the greedy algorithm: one chooses, at each stage, the largest circle that fits in the space.

61. A Short Bisector

Suppose first that the segment goes from AB to AC, as illustrated. The area of $\triangle ABC$ is 6 and the segment must cut this area in half, so the area of $\triangle APQ$ must be 3. Since $\triangle APQ$ has base x and altitude $y \sin\alpha$, the area of $\triangle APQ$ is $\frac{1}{2}xy\sin\alpha$, so we must have $xy = 6/\sin\alpha$.

By the law of cosines, the square of the length of PQ is

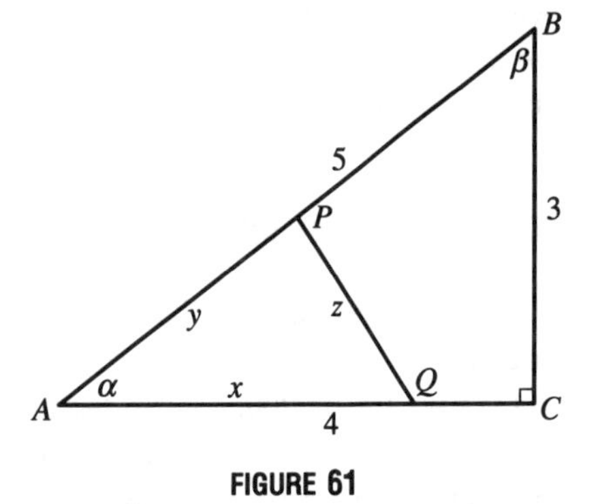

FIGURE 61

$$z^2 = x^2 + y^2 - 2xy\cos\alpha = (x-y)^2 + 2xy(1-\cos\alpha) = (x-y)^2 + \frac{12(1-\cos\alpha)}{\sin\alpha}.$$

Clearly, this is minimized when $x = y$, and the minimum value is

$$z^2 = \frac{12(1-\cos\alpha)}{\sin\alpha} = \frac{12(1-\frac{4}{5})}{\frac{3}{5}} = 4,$$

or $z = 2$. This minimum is achieved when $x^2 = xy = 6/\sin\alpha = 10$, so $x = y = \sqrt{10} \approx 3.162$.

Similarly, if the segment goes from AB to BC then z^2 cannot be smaller than $12(1-\cos\beta)/\sin\beta$, which equals 6, and if the segment goes from AC to BC then z^2 cannot be smaller than $12(1-\cos 90°)/\sin 90°$, which equals 12. Thus, the segment from AB to AC is the shortest one.

Problem 61.1. It is possible to divide the triangle in half with a cut whose length is less than 2, if we use a cut that is not a straight line. Can you do it?

Notes. This problem appeared in [Kon]. The optimal solution to Problem 61.1 can be found in [Wie]. For a discussion of this and related problems, see [CFG, problem A26].

62. The Pizza Problem

It turns out that the equal-area property holds if and only if the number of chords is even and greater than or equal to 4 (equivalently, the number of pizza slices is 8, 12, 16, ...). We begin our discussion by proving the negative result in the case of three chords. In what follows, P denotes the point common to all the chords.

In the case of three chords the areas are not equal. This can be easily seen by placing P on the pizza-edge (Figure 62), with one of the chords tangent at P. Elementary geometry yields inequality (the central region has area $\pi/3 + \sqrt{3}/2$, assuming radius 1); by continuity, inequality remains valid when the point is moved a little away from the edge.

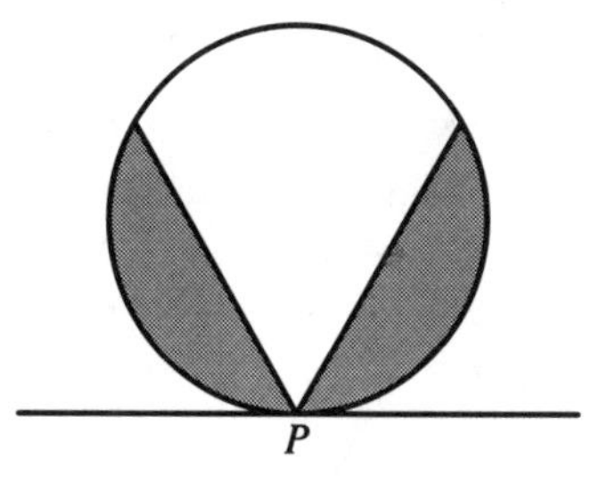

FIGURE 62

For 4 chords the equal-area result is true, a very pretty theorem first discovered by L. J. Upton in 1968 [Upt]. Figure 63 shows a proof by dissection, discovered by L. Carter and S. Wagon [CW] with the help of the geometry software *CABRI*. The key to the dissection is the octagon (dotted lines) formed by the 8 points $(\pm x, \pm y)$ and $(\pm y, \pm x)$, where $P = (x, y)$. It is easily checked that it works regardless of the position of P. Similar proofs in

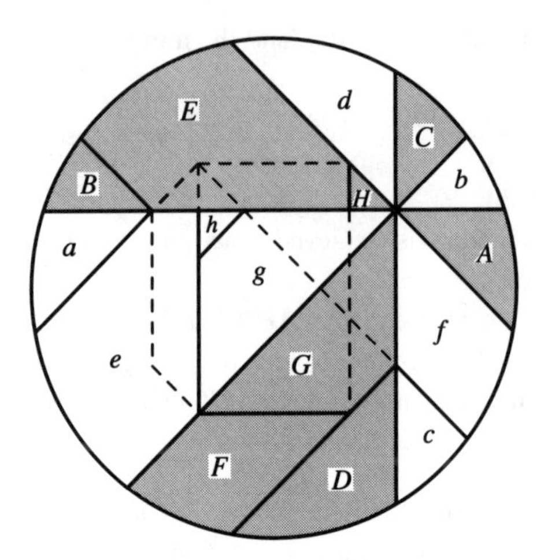

FIGURE 63
A dissection solution to the pizza problem.

the case of 6 or 8 chords have been found by Allen Schwenk (Western Michigan University).

However, it is more illuminating to use calculus and polar coordinates, as was done by Jörg Härterich [Rab4] and, in essence, by Michael Goldberg [Upt]. First we need an auxiliary lemma.

Lemma. *Suppose two chords AC and BD in a circle of radius R intersect at P making four right angles and giving rise to the four line segments PA, PB, PC, PD with lengths a, b, c, d, respectively. Then $a^2 + b^2 + c^2 + d^2 = 4R^2$.*

Proof. Make a right triangle as illustrated in Figure 64; Pythagoras then yields $(2R)^2 = (c-a)^2 + (b+d)^2$. A similar construction in the other direction tells us that $(2R)^2 = (a+c)^2 + (b-d)^2$. Expanding and combining yields the result.

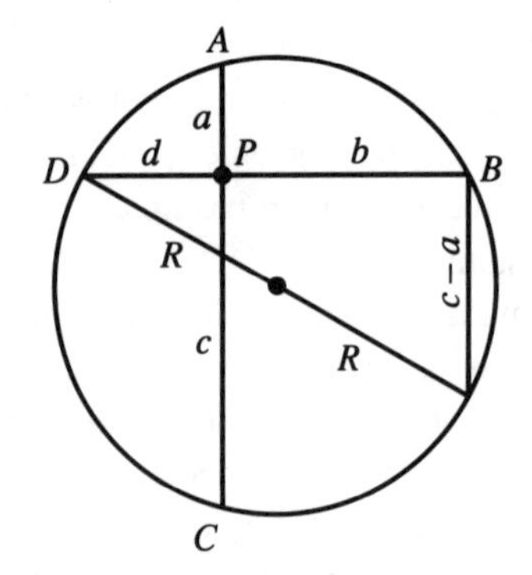

FIGURE 64

Now return to the main problem. We assume in all that follows that the circle's radius is 1. Choose polar coordinates with the origin at P and the polar axis lying along one of the cuts; let the first slice counterclockwise from the polar axis be colored black.

Let $r(\theta)$ denote the distance from P to the circle at angle θ with the polar axis. The area of the black region is then

$$\frac{1}{2}\left(\int_0^{\pi/4} r^2(\theta)\,d\theta + \int_{\pi/2}^{3\pi/4} r^2(\theta)\,d\theta + \int_{\pi}^{5\pi/4} r^2(\theta)\,d\theta + \int_{3\pi/2}^{7\pi/4} r^2(\theta)\,d\theta\right).$$

Some substitutions transform this to

$$\frac{1}{2}\int_0^{\pi/4} r^2(\theta) + r^2\left(\theta + \tfrac{\pi}{2}\right) + r^2(\theta+\pi) + r^2\left(\theta + \tfrac{3\pi}{2}\right)\,d\theta,$$

which, by the lemma, is simply $\frac{1}{2}\int_0^{\pi/4} 4\,d\theta$ or $\pi/2$, as required.

Notes. The equal-area result holds if the number of chords is one of $4, 6, 8, 10, \ldots$. This is because the lemma holds if the number of chords is 2 or more. The easiest way to prove this is analytically: assume $P = (x, 0)$ and let θ_0 denote the angle between the x-axis and the first chord in the counterclockwise direction. Use the law of cosines to get an explicit expression for $r(\theta)$ in terms of x and then use various identities that cause extreme simplification in the desired expression, $\sum_{i=1}^{2n} r\left(\theta_0 + (2\pi i/2n)\right)^2$; alternatively, there is a geometric proof (see [Fuk]). The proof of the two-chord case of the lemma given here is due to Emil Slowinski (chemist, Macalester College).

In the remaining cases, when the number of chords is $1, 2, 3, 5, 7, 9, 13, \ldots$, the equal-area result fails. The case $n = 2$ is easily treated as a special case. If n is odd we can take an abstract approach (due to Don Coppersmith of IBM) as follows. It suffices, by continuity, to take P on the boundary of the circle and one of the chords to be the tangent at P. Then the black area can be expressed in terms of π and algebraic numbers in such a way that its equality with $\pi/2$ would yield an algebraic relationship for π, in contradiction to π's transcendence (details omitted). But this transcendental sledgehammer can be avoided by showing explicitly that the difference between the black and white areas is $\pi/n - \tan(\pi/n)$, which is nonzero.

To do this, assume that n is odd, that the first pizza slice, reading counterclockwise, is black, and that θ denotes π/n. Then observe that the arcs at the ends of the pizza slices all make angles of 2θ at the center (Figure 65). The first slice, being simply a sector minus a triangle, has area $\frac{1}{2}2\theta - \frac{1}{2}\sin(2\theta)$, or $\theta - \frac{1}{2}\sin(2\theta)$. The next black slice is the difference between two regions of the form of the first slice, but with a central angle three and two times as large, respectively; thus this slice has area

$$\left(3\theta - \frac{1}{2}\sin(6\theta)\right) - \left(2\theta - \frac{1}{2}\sin(4\theta)\right) = \theta - \frac{1}{2}\sin(6\theta) + \frac{1}{2}\sin(4\theta).$$

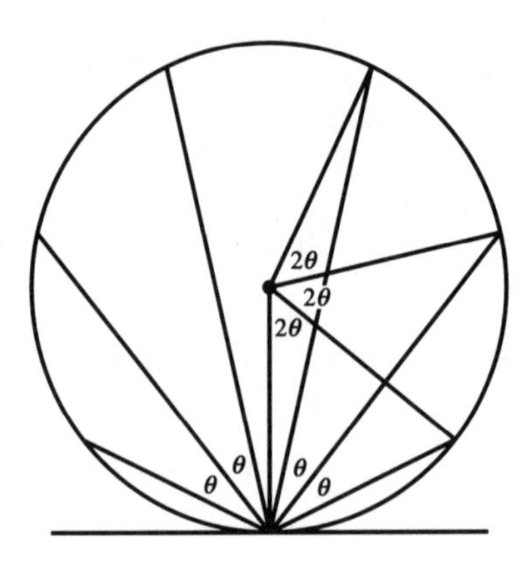

FIGURE 65

Continuing in this way, we see that the total area of the black region is

$$\frac{(n+1)\theta}{2}+\frac{1}{2}\sum_{k=1}^{n}(-1)^k\sin(2k\theta).$$

This can be simplified. We let *Mathematica* work on the summation.

```
Needs["Algebra`SymbolicSum`"]                (*unnecessary in version 3)
SymbolicSum[(-1)^k Sin[2 k theta], {k, 1, n}]
Cos[n Pi/2 + theta + n theta] Sec[theta] Sin[n Pi/2 + n theta]
```

This identity can be verified by induction. Now, if θ is replaced by π/n, this formula becomes $\cos(n\pi/2+\pi/n+\pi)\sec(\pi/n)\sin(n\pi/2+\pi)$, which standard trig identities reduce to

$$\frac{(-1)^n-1}{2}\tan\frac{\pi}{n}.$$

This summation formula is true for all integers n, but if n is odd it reduces to $-\tan(\pi/n)$ and the black area is

$$\frac{(n+1)\pi}{2n}-\frac{\tan(\pi/n)}{2}.$$

Since the black area minus the white area is twice the black area less π, the difference in areas simplifies to $\pi/n-\tan(\pi/n)$, as claimed.

Problem 62.1. If a square pizza is divided into 8 pieces by cuts through a point that are parallel to the sides and diagonals of the square, and if the pieces are colored alternately, then the black area equals the white area.

For further investigation. There are many variations to the pizza problem that seem worthy of investigation. Allen Schwenk suspects that there are dissection proofs for all the positive cases and he has found such proofs when n, the number of chords, is 6 or 8. Is there a uniform way of describing such dissection proofs for arbitrary even n greater than 4? As far as negative results go, suppose that n, the number of chords, is one of $3, 5, 7, \ldots$. Numerical computations by Larry Carter and Stan Wagon provided evidence for the conjecture [CDW] that the color of the region containing the center of the circle defines the color of greater area (respectively, lesser area) if n has the form $4k - 1$ (resp., $4k + 1$). John Duncan (University of Arkansas, Fayetteville) has proved this conjecture when $n = 3$, and the $n = 5$ and $n = 7$ cases have been proved by Paul Deiermann and Rick Mabry (Louisiana State University, Shreveport; see [CDW]) A consequence of this conjecture is that inequality of area always holds in these negative cases, unless one of the chords is a diameter of the circle. A comprehensive summary of results in this area may be found in [Rab5].

63. More Pizza!

Let A_i denote the points on the circle, in counterclockwise order. The result is clear if P is the center of the circle, so it is sufficient to show that the black area is independent of the placement of P. If we look at the polygon obtained by connecting the A_i in sequence, then the black area is a sum of triangles and pieces with circular boundaries. (If P lies outside this polygon, then we can, if necessary, switch colors to guarantee the truth of the previous sentence.) The curvy pieces are clearly independent of P. To study the triangular areas, use vectors: let $\mathbf{P}$ be the vector from the center of the circle to P and let $\mathbf{A}_i$ be the vector from the center to A_i. View these vectors as being horizontal vectors in 3-space! Then, because the length of the cross product of two vectors in 3-space is the area of a parallellogram, the triangular part of the black area is

$$\tfrac{1}{2}\Big(\big|(\mathbf{A}_1-\mathbf{P})\times(\mathbf{A}_2-\mathbf{P})\big|+\big|(\mathbf{A}_3-\mathbf{P})\times(\mathbf{A}_4-\mathbf{P})\big|+\cdots+\big|(\mathbf{A}_{n-1}-\mathbf{P})\times(\mathbf{A}_n-\mathbf{P})\big|\Big).$$

But by the right-hand rule all the cross products point upward. Thus the expression is the length of a single vector:

$$\tfrac{1}{2}\big|(\mathbf{A}_1-\mathbf{P})\times(\mathbf{A}_2-\mathbf{P})+(\mathbf{A}_3-\mathbf{P})\times(\mathbf{A}_4-\mathbf{P})+\cdots+(\mathbf{A}_{n-1}-\mathbf{P})\times(\mathbf{A}_n-\mathbf{P})\big|.$$

Now expand, using the rules: $\mathbf{P} \times \mathbf{P} = 0$, $\mathbf{A} \times \mathbf{B} = -\mathbf{B} \times \mathbf{A}$. The area then becomes:

$$\tfrac{1}{2}\Big|\mathbf{P} \times (\mathbf{A}_1 - \mathbf{A}_2 + \mathbf{A}_3 - \mathbf{A}_4 + \cdots + \mathbf{A}_{n-1} - \mathbf{A}_n) + \mathbf{A}_1 \times \mathbf{A}_2 + \mathbf{A}_3 \times \mathbf{A}_4 + \cdots + \mathbf{A}_{n-1} \times \mathbf{A}_n\Big|.$$

But $\mathbf{A}_1 + \mathbf{A}_3 + \cdots + \mathbf{A}_{n-1} = 0 = \mathbf{A}_2 + \mathbf{A}_4 + \cdots + \mathbf{A}_n$, and the second summand is independent of P. This concludes the proof.

Notes. This beautiful proof by algebra is due to Murray Klamkin. In fact, there is a single theorem that generalizes both pizza results: Klamkin's version and that of Problem 62. This was discovered by Stanley Rabinowitz. Suppose n is a multiple of 4 and let Q be an arbitrary point inside a circle. Draw n equiangular chords through Q and let A_i be the intersection points of these chords with the circle. Now let P be any point inside the polygon determined by the A_i and connect P to each A_i to get $2n$ pizza pieces. Color the pieces alternately black/white. Then the black area equals the white area. Note that if $P = Q$ this result becomes the standard pizza theorem of Problem 62 while if Q is the center of the circle then the result becomes the Klamkin pizza problem (Problem 63). A proof of the generalization may be found in [Rab5].

7.7 Miscellaneous

64. Nearest Neighbors

Suppose point A is connected to six other points. The lines from A to these six points form six angles at A that sum to 360°. Let $\angle BAC$ be the smallest of these angles, where B and C are two of the six points connected to A, and let θ be the measure of $\angle BAC$. Note that $\theta \leq 60°$. Let b and c be the distances from A to B and C respectively, and assume without loss of generality that $b > c$. Then by the law of cosines, the square of the distance from B to C is $b^2 + c^2 - 2bc\cos\theta \leq b^2 + c^2 - 2bc\cos 60° = b^2 + c^2 - bc = b^2 + c(c - b) < b^2$, since $c - b < 0$. Thus the distance from B to C is smaller than the distance from B to A, so A is not the closest point to B, and the distance from A to C is smaller than the distance from A to B, so B is not the closest point to A. Therefore A and B should not have been connected.

Problem 64.1. Give an example showing that a point can be connected to five others.

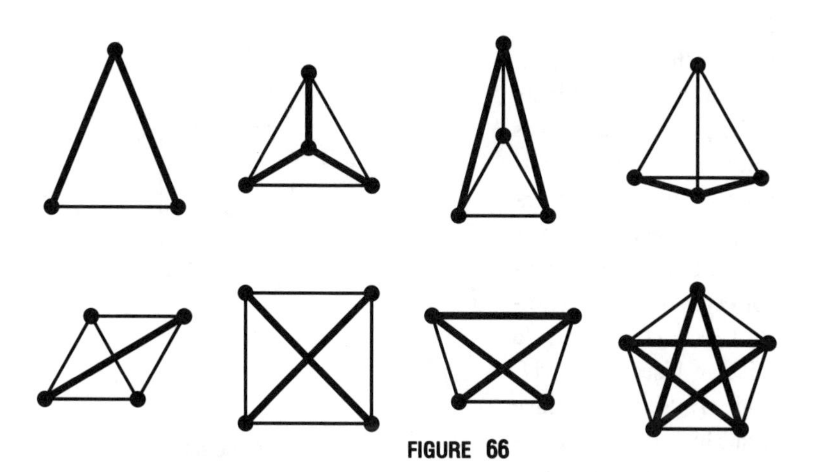

FIGURE 66

65. Two-Distance Sets

The only 2-distance sets are the arrangements of points shown in Figure 66. In each of the figures, the thin lines have one length and the thick lines have another.

Problem 65.1. Prove that there are no other 2-distance sets.

Note. For an extensive discussion of 2-distance sets, see [ES].

66. An Almost-Symmetric Hexagon

Extend the sides and label the intersection points as illustrated. The given $120°$ angles imply that the resulting large triangle and the three smaller triangles are all equilateral. Therefore $PA+AF+FQ=QE+DE+DR$, which, because of the smaller triangles, yields $AB+AF+EF=EF+DE+CD$, or $AB-DE=CD-FA$. The other equation is similar.

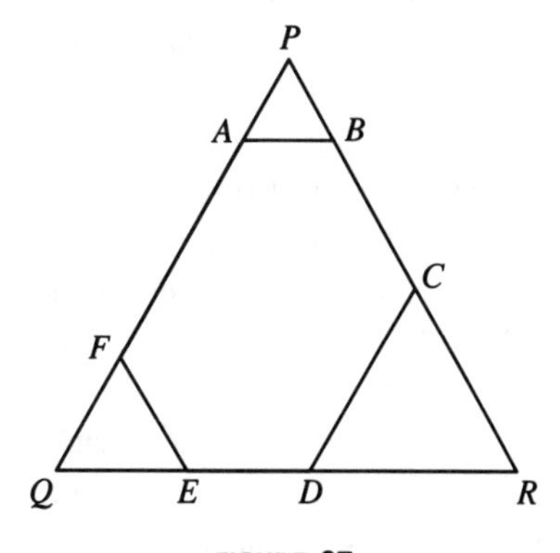

FIGURE 67

67. An Eight-Point Arrangement

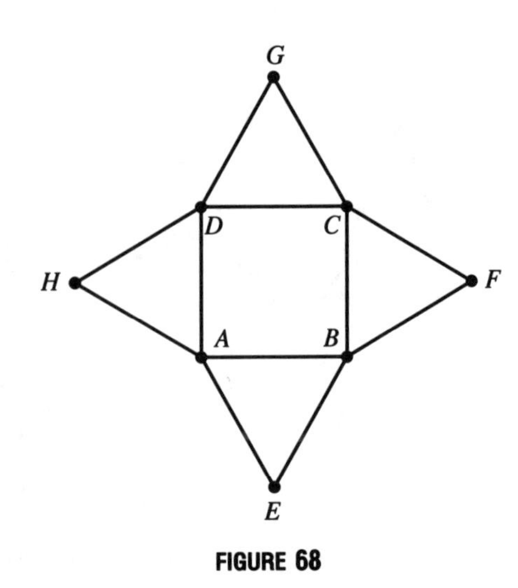

FIGURE 68

To create the required arrangement, draw a square with an equilateral triangle on each side (see Figure 68). Note that, since the angles in the triangles and square are 60° and 90° respectively, if GD is extended through D then it bisects $\angle ADH$, and therefore it is the perpendicular bisector of AH. It follows that $AG = HG = FG$, so the perpendicular bisector of AF passes through G. Similarly, since $AB = FB$, this bisector also passes through B. It should now be clear that the perpendicular bisector of the segment joining any two points in the arrangement passes through exactly two of the points.

Notes. It is not known whether there is an arrangement of more than eight points with the same property. For more on this and related problems see [CFG, problem F10].

68. Distances Determined by Five Points

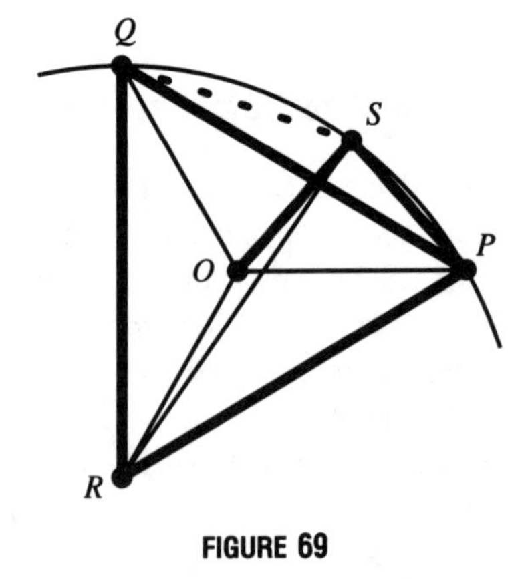

FIGURE 69

Figure 69 gives a solution. Start with an equilateral triangle PQR with center O. Let S be the indicated intersection point of the perpendicular bisector of OP with the circle centered at R and passing through P and Q. Then $PR = QR = SR = PQ$, $OP = OQ = OR$, $SP = SO$, and QS is distinct from all other distances. To prove that these families of distances are distinct, choose units so that $OP = 1$; straightforward computations show that the distances are $PR = \sqrt{3}$, $OP = 1$, $PS = \sqrt{3 - \sqrt{6}}$, and $QS = \sqrt{6 - 2\sqrt{6}}$. It is easy to see that no four of these points lie on a circle; and because $SO + OR > SR$, which reduces to $2\sqrt{6} < 5$, no three of these points lie on a line.

Notes. Call a set of n points in a plane an EPS(n) set (Erdős point set) if one distance occurs once, another occurs twice, and so on up to the most common distance which occurs $n-1$ times, and no three points of the set lie on a line and no four on a circle. In 1983 Paul Erdős asked: For which n do EPS(n) sets exist? The EPS(5) exhibited above is due to Carl Pomerance. It is now known that EPS(n) sets exist if n is 2, 3, 4, 5, 6, 7, or 8 (see [Liu2, Pal]). A picture of Palásti's EPS(8) is presented in color plate 3 following page 16. H. Harborth and L. Piepmeyer [HP] have investigated EPS(n) sets in which all distances are integers; an integral EPS(4) set exists, but there are no integral EPS(5) sets. In the same paper, Harborth and Piepmeyer showed that there are precisely 89 different EPS(5) sets.

Problem 68.1 [Liu2]. Find all EPS(4) sets.

For Further Investigation. Does an EPS(9) set exist?

69. What's the Angle?

Let $\alpha = \angle DAE$. Since $\triangle ABC$ is isosceles, $\angle ACB = \alpha$ too. Therefore $\angle ABC = 180° - 2\alpha$, so $\angle CBD = 2\alpha$. But $\triangle BCD$ is also isosceles, so $\angle BDC = 2\alpha$, $\angle BCD = 180° - 4\alpha$, and $\angle DCE = 3\alpha$. Since $\triangle CDE$ is isosceles, $\angle AED = 3\alpha$. By symmetry, $\angle ADE = 3\alpha$ too. Now adding the angles in $\triangle ADE$ we get $\alpha + 3\alpha + 3\alpha = 180°$, so $\alpha = 180°/7 \approx 25.7°$.

Problem 69.1. Generalize!

70. Find the Common Point

Choose C on the bisector of $\angle O$ so that it lies on AB (see Figure 70). It suffices to prove that the position of C is a function of k only. Choose D on OB so that DC is parallel to OA. Then $\angle DOC = \angle DCO$ and $\angle BCD = \angle BAO$, so $DC = DO$ and $\triangle CDB$ is similar to $\triangle AOB$. This yields:

$$\frac{OB}{OD} = \frac{BD + OD}{OD} = \frac{BD}{DC} + 1 = \frac{OB}{OA} + 1.$$

Dividing by OB gives $1/OD = k$. But, because of isosceles triangle ODC, the position of D determines the position of C. Hence C is completely determined by the value of k.

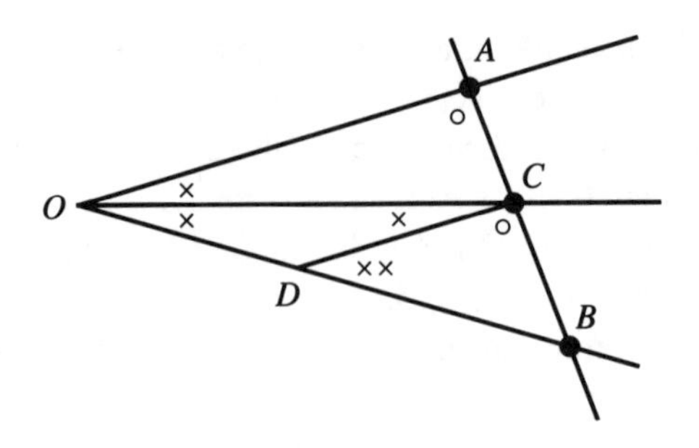

FIGURE 70

71. A Shady Garden Wall

Yes. Here's one way to do it.

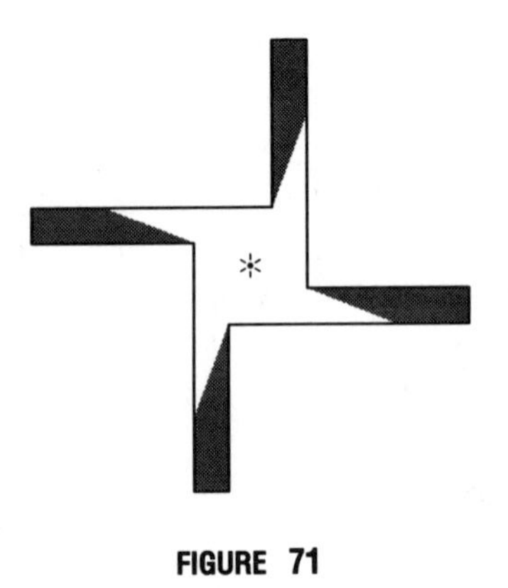

FIGURE 71

Problem 71.1. Is it possible to make a polygonal garden wall and position a light source *outside* it so that part or all of every wall is in shadow?

Note. Although this problem is not too difficult, there are many illumination problems that are exceedingly difficult, especially when the walls are mirrored and reflections are allowed. See [KW, section 1] and [Tok].

72. A Square Peg in a Nonround Hole?

No, there are other possible shapes for the hole. Suppose each side of the square peg has length $\sqrt{2}$. Then if the peg is rotated 90° around its center, the four corners

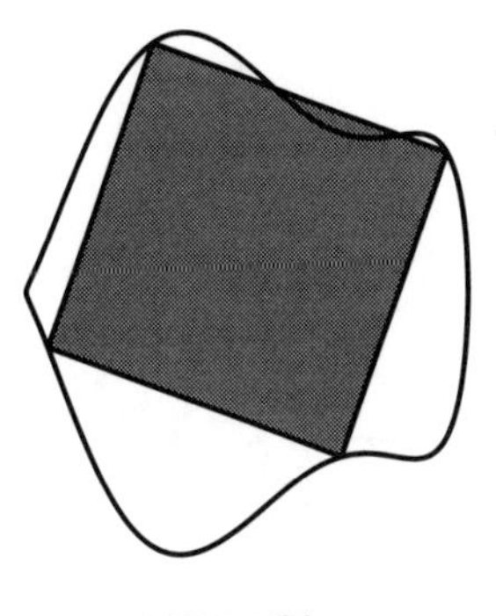

FIGURE 72

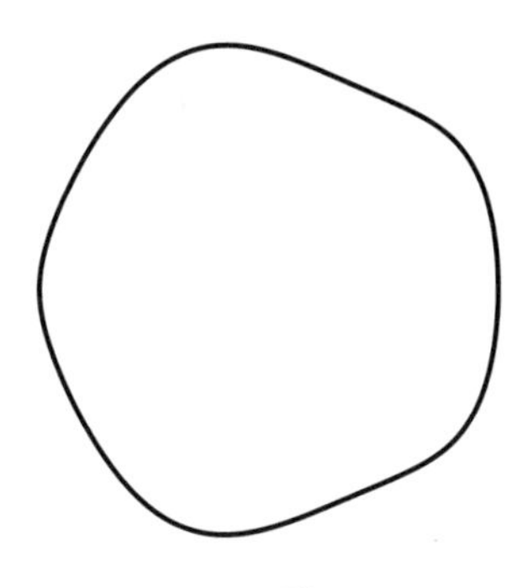

FIGURE 73

of the peg will trace out four quarters of a circle of radius 1, and this circle is one possible shape for the hole. We can generate other possible shapes by letting the peg "wobble" as it rotates. Since we want the wobble to be completed in a 90° rotation, a reasonable choice would be to have the center of the square be at the point $(0, \varepsilon \sin 4\theta)$ after a rotation through an angle θ, where ε is a parameter that determines the size of the wobble. It is easy now to write down parametric equations for the curve traced out by the corners of the square as it rotates: $x = \cos\theta$, $y = \varepsilon \sin 4\theta + \sin\theta$, $0 \le \theta \le 2\pi$.

Note that for every angle ρ, the points corresponding to $\theta = \rho$, $\theta = \rho + \frac{\pi}{2}$, $\theta = \rho + \pi$, and $\theta = \rho + \frac{3\pi}{2}$ form the corners of a square with side-length $\sqrt{2}$, with a diagonal of the square making an angle ρ with the horizontal. Thus, if this curve forms the boundary of the hole, then no matter how the peg is oriented, it can be positioned over the hole so that all four corners touch the edge of the hole.

However, the peg may still not fit through the hole in some orientations, as Figure 72 illustrates! In this figure, we have used the value $\varepsilon = \frac{1}{5}$.

One way to solve this problem is to choose ε small enough that the hole is convex. This guarantees that if the corners of the peg are on the edge of the hole, then the rest of the peg is inside the hole. As we will show, if $\varepsilon = \frac{1}{15}$ then the hole is convex. Figure 73 shows the hole with $\varepsilon = \frac{1}{15}$.

One way to confirm that the hole in Figure 73 is convex is to convert the parametric equations defining the boundary of the hole to an equation giving y as a function of x. Since $\sin 4\theta = 2\sin 2\theta \cos 2\theta = 4 \sin\theta \cos\theta (2\cos^2\theta - 1)$, we have

$$\begin{aligned} y &= \tfrac{1}{15} \sin 4\theta + \sin\theta \\ &= \tfrac{4}{15} \sin\theta \cos\theta (2\cos^2\theta - 1) + \sin\theta \\ &= \sin\theta \big[1 + \tfrac{4}{15}(2\cos^3\theta - \cos\theta)\big]. \end{aligned}$$

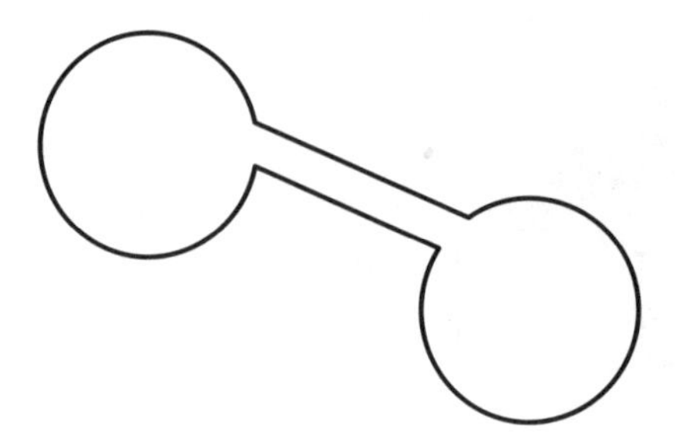

FIGURE 74

Now because $x = \cos\theta$, $\sin\theta = \pm\sqrt{1-x^2}$, so we have

$$y = \pm\sqrt{1-x^2}\left[1 + \tfrac{4}{15}(2x^3 - x)\right],$$

where of course the positive and negative square roots give the top and bottom halves of the hole respectively. It is now a messy but routine calculus exercise to verify that the curve defining the top half of the hole is concave down everywhere, and the bottom is concave up everywhere. Thus the hole is convex.

Problem 72.1. What is the largest value of ε for which the hole is convex?

Another possible solution, which might be considered cheating, is the hole shown in Figure 74. The square peg just fits through each of the circular parts of the hole. Thus, for each orientation of the peg there are two ways to fit it through the hole, and for at least one of those orientations all four corners touch. But for some orientations there is a way to put the peg through the hole so that one corner doesn't touch.

For Further Investigation. What happens with pegs of other shapes?

73. A Most Elementary Fact

Let B', C', D', E' lie on AC, AC, AE, AE, respectively so that $DD' = DF$, $BB' = BF$, $EE' = EF$, $CC' = CF$. Then $AD' = AB'$; moreover, the problem is reduced to showing that $AE' = AC'$. Draw angle bisectors at A, B, and D as in Figure 75 and let α, β, and γ denote the measures of the resulting half-angles.

By the exterior angle theorem $\angle ACD = 2\beta - 2\alpha$, and so $\angle B'C'F = \beta - \alpha$ since $\triangle CC'F$ is isosceles. And also $\angle B'D'F = \beta - \alpha$ since the angle equals $(90° - \alpha) - (90° - \beta)$ (note the 90° angles marked in Figure 75). Therefore $B'D'C'F$ is a cyclic quadrilateral. Similarly, $B'D'E'F$ is cyclic. Therefore the

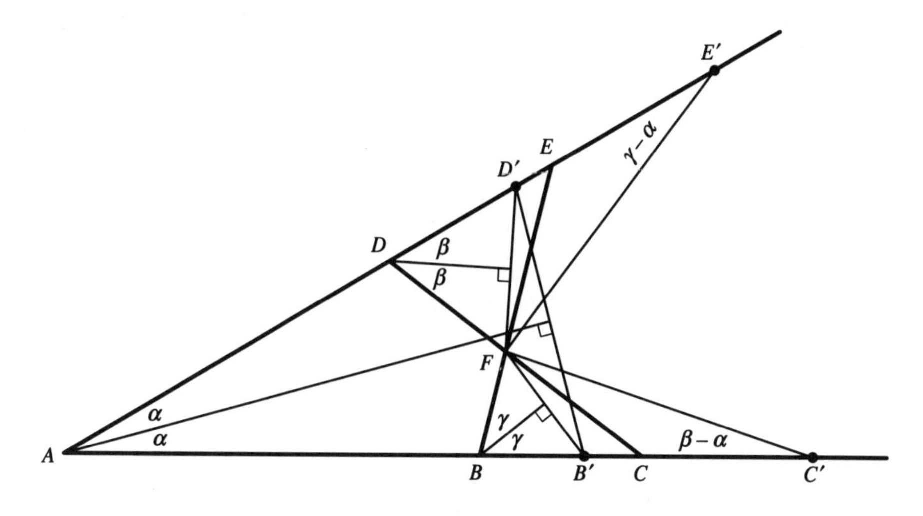

FIGURE 75

five points B', D', C', E', and F lie on a common circle. Using the exterior angle theorem again, $\angle B'FC' = (90° - \gamma) - (\beta - \alpha)$ and $\angle D'FE' = (90° - \beta) - (\gamma - \alpha)$. So chords $B'C'$ and $D'E'$ subtend equal angles at the circle and so $B'C' = D'E'$. Therefore $AE' = AC'$, which suffices.

Notes. This theorem is due to M. L. Urquhart (1902–1966), who discovered it while making some investigations into the geometry of space-time. He called it the "most elementary theorem" since it involves only the concepts of straight lines and distance [Ell]. The proof given here is due to John Duncan (University of Arkansas). Various other proofs and comments can be found in [Eus, Ped, Sok1, Sok2, Sze, Wil]; the proof in [Sok2] is noteworthy in that it makes no reference to circles.

CHAPTER 8

Number Theory

8.1 Prime Numbers

74. No Primes Here

Let a_{ij} denote the entry in row i, column j of the array. The first row is an arithmetic progression with first element 4 and common difference 3, so, for every j, $a_{1j} = 3j+1$. Similarly, $a_{2j} = 5j+2$. By the symmetry of the table, $a_{i1} = 3i+1$ and $a_{i2} = 5i+2$, so row i must be an arithmetic progression with first entry $3i+1$ and common difference $2i+1$. Therefore $a_{ij} = i + (2i+1)j = 2ij + i + j$.

The statement to be proved is equivalent to the statement that if we double all the numbers in the array and then add 1 to each entry, then the resulting array will contain precisely the odd composite numbers. But if we double the numbers in the array and add 1, then the entry in row i, column j of the new array will be

$$2(2ij + i + j) + 1 = 4ij + 2i + 2j + 1 = (2i+1)(2j+1),$$

so the desired conclusion is clear.

Note. This problem appeared in [Zai].

75. Squares, Primes, and Squares Plus Primes

The following lemma settles both issues, since there are infinitely many n for which $2n-1$ is prime, and infinitely many for which it is not (for the latter let n have the form $3m-1$).

Lemma. *The integer n^2 is the sum of a perfect square and a prime if and only if $2n-1$ is prime.*

Proof. If $2n-1$ is prime, then, since $n^2 = (n-1)^2+2n-1$, n^2 is a square plus a prime. Conversely, if $n^2 = m^2+p$, p prime, then $p = n^2-m^2 = (n-m)(n+m)$, so $n-m = 1$ and $n+m = p$. But this means $p = n+(n-1) = 2n-1$.

76. Not Goldbach's Conjecture

Look at even numbers of the form $m = 6n+k$, and consider the 3 possibilites: $k = 0, 2,$ or 4. If $k = 0$, then $m = 6n = 9+3(2n-3)$, which works if $2n-3 > 1$, or $m > 12$. If $k = 2$, then $m = 6n+2 = 35+3(2n-11)$, which works provided $2n-11 > 1$, or $m > 38$. Finally, if $k = 4$, then $m = 6n+4 = 25+3(2n-7)$, which works if $2n-7 > 1$, or $m > 28$. This analysis tells us that every even integer larger than 38 is a sum of two odd composites. But it is easy to check that 38 is not such a sum, so 38 is the largest even integer that is not a sum of two odd composites.

77. They're There, but Where?

Let $p(n) = n^2+n+41$ and let $N = p(0)p(1)\cdots p(39)$. Then

$$\begin{aligned} p(N+k) &= (N+k)^2+(N+k)+41 \\ &= N^2+2kN+k^2+N+k+41 \\ &= N^2+2kN+N+k^2+k+41 \\ &= N(N+2k+1)+p(k). \end{aligned}$$

But $p(k)$ divides N if $k = 0, 1, 2, \ldots, 39$, whence $p(k)$ divides $p(N+k)$ in these cases. Moreover, $N(N+2k+1) > 0$ because $N > 0$, and so $p(N+k) > p(k)$, proving that $p(N+k)$ is not prime.

Notes. The proof above, with very little change, shows that for any nonconstant polynomial $p(n)$ with integer coefficients and any M, there are M consecutive integer values of n for which $p(n)$ is composite. To learn more about Euler's curious polynomial from a modern viewpoint, see [Fen] or [Fla, Appendix A].

78. A Pascal Pattern

The solution is short and sweet, but perhaps not so easy to find. Suppose the two entries are $\binom{n}{i}$ and $\binom{n}{j}$ where $0 < i < j < n$. Then $\binom{n}{i} > \binom{j}{i}$ and the result follows immediately from the easily verified identity,

$$\binom{n}{j}\binom{j}{i} = \binom{n}{i}\binom{n-i}{j-i}.$$

Note. This problem is due to P. Erdős and G. Szekeres [ESz].

8.2 Digits

79. 1011 and All That

Let s_i be the sum of the first i members of the set $\{1, 10, 100, 1000, \ldots\}$. There are only N possibilities for the remainder after division by N. This means that among the first $N+1$ sums s_i, there are two of them, s_j and s_{j+k}, that leave the same remainder. Then $s_{j+k} - s_j$, which looks like $111\cdots111000\cdots000$, is divisible by N.

80. Only Ones

Let U_n denote the number consisting of n 1s; then $U_n = (10^n - 1)/9$. The result now follows from the following assertion: If U_n is divisible by n, then U_{3n} is divisible by $3n$. For this implies that U_n is divisible by n whenever n is a power of 3. To prove the assertion assume $U_n = kn$. Then

$$\begin{aligned} U_{3n} &= (10^{3n} - 1)/9 = (10^n - 1)(1 + 10^n + 10^{2n})/9 \\ &= U_n(1 + 10^n + 10^{2n}) = kn(1 + 10^n + 10^{2n}). \end{aligned}$$

Because 10 has the form $3s+1$, so do 10^n and 10^{2n}, and therefore $1 + 10^n + 10^{2n}$ is a multiple of 3, as desired.

Problem 80.1. Show that if U_n is divisible by n, then U_{U_n} is divisible by U_n.

Problem 80.2. Problem 80.1 and the solution to problem 80 contain two methods for generating new values of n that work from old ones. These two methods, starting from 1, generate the list $\{1, 3, 9, 27, 81, 111, 243, 333, 729, 999, \ldots\}$. Are there any integers n that are not in this list, but for which n divides U_n?

Problem 80.3. If n divides U_n (which is $111\ldots1$), then of course n divides $10^n - 1$ (which is $9999\ldots99$). Are there any values of n that divide $10^n - 1$, but do not divide U_n?

81. Getting Ten Digits

The assertion is true. Given n, let d be the number of digits in its base-10 representation and consider the integers between $1234567890 \cdot 10^d + 1$ and $1234567890 \cdot 10^d + n$, inclusive. Each integer in this sequence contains all ten digits; since any sequence of n consecutive integers contains one that is divisible by n, we are done.

82. Reversing Multiples

Suppose the number to be reversed is $1000p + 100q + 10r + s$, where each of p, q, r, and s is one of the digits 0–9 and $p \neq 0$. We must solve the equation

$$4(1000p + 100q + 10r + s) = 1000s + 100r + 10q + p.$$

The left side of the preceding equation is even, so p must be even, and the right side is at most 9999, so $p \leq 2$. Thus $p = 2$.

Since $p = 2$, the left side of our equation is at least 8000, and therefore $s \geq 8$. But if $s = 9$ then the last digit of the left side of the equation would be a 6, whereas the last digit of the right side is a 2. Therefore $s = 8$. Substituting the known values for p and s we get

$$4(2000 + 100q + 10r + 8) = 8000 + 100r + 10q + 2,$$

or $13q + 1 = 2r$. Since $2r$ is even and no greater than 18, q must be 1, which yields $r = 7$. So the only four-digit number that is reversed by multiplication by 4 is 2178.

Problem 82.1. Are there numbers that are reversed by multiplication by 2, 3, 5, 6, 7, or 8?

Problem 82.2. Is 1089 the only number that is reversed by multiplication by 9?

For Investigation. Try varying the digit-length of the number to be reversed. Try bases other than base 10.

83. Can You Beat a Billion?

We note two key facts:

(a) If the two-digit number $10a + b$ is one of the factors of the maximum product then $a > b$, for otherwise $10b + a$ is larger.

(b) If $a_1 > a_2$, then

$$(10a_1 + b_1)(10a_2 + b_2) - (10a_1 + b_2)(10a_2 + b_1) = 10(a_1 - a_2)(b_2 - b_1),$$

which is positive if and only if $b_2 > b_1$.

Now, suppose $ab \cdot cd \cdot ef \cdot gh \cdot ij$ is the desired largest product. By (a), we may assume $a = 9$. Then, by (b), d, f, h, j are all larger than b, which means that b is not one of $\{9, 8, 7, 6, 5\}$. Then 8 must be one of the tens digits, so assume the number is $9b \cdot 8d \cdot ef \cdot gh \cdot ij$. Again by (b), each of f, h, and j are greater than d, whence d is not in $\{9, 8, 7, 6, 5\}$ and we may assume e is 7. Arguing in this way brings our number to the form $9b \cdot 8d \cdot 7f \cdot 6h \cdot 5j$. But b is the smallest of the units digits, so b is 0, with similar arguments showing that d is 1, and so on. The desired maximum is therefore $90 \cdot 81 \cdot 72 \cdot 63 \cdot 54$, or 1,785,641,760.

84. A Classy Social Security Number

Suppose $a_1a_2a_3a_4a_5a_6a_7a_8a_9$ has the desired property. Then we get immediately that $a_5 = 5$ and a_2, a_4, a_6, and a_8 are even. By the well-known divisibility-by-3 test, $a_1 + a_2 + a_3$ and $a_4 + 5 + a_6$ must each be divisible by 3. It follows that $a_4a_5a_6$ is either 258 or 654 (the other two choices, 852 and 456, are eliminated by the oddness of a_3, which would cause a failure of the divisibility-by-4 test).

The 258 choice leads only to the possibilities 147258369, 147258963, 741258369, 741258963, 369258147, 369258741, 963258147, 963258741. But the sequences containing 836, 814, or 874 are forbidden by the divisibility-by-8 condition, while the remaining two candidates fail the 7-test.

So the only possibility is $***654***$. The 8-test tells us that the form must be $***654x2*$ where x is 3 or 7, and this yields the eight possibilities: 987654321, 789654321, 381654729, 183654729, 981654327, 189654327, 981654723, and 189654723. Only one of these, 381654729, survives the divisibility-by-7 test, so it's the unique solution. Note that the solution must satisfy a 9-divisibility test too, since the sum of the digits is 45.

85. Three Repeated Digits in Two Bases

Suppose a and b are the bases for the representations sought, with $a < b$. Then there are positive integers x and y, with $x < a$ and $y < b$, such that

$$xa^2 + xa + x = yb^2 + yb + y,$$

and so

$$x(a^2 + a + 1) = y(b^2 + b + 1)$$

and $x > y$. Let a' and b' be the two quotients after division of a^2+a+1, b^2+b+1, respectively, by $\gcd(a^2 + a + 1, b^2 + b + 1)$; then $\gcd(a', b') = 1$, and $xa' = yb'$ so that b' divides x and a' divides y. For a minimum, we may choose $x = b'$ and $y = a'$.

Now, the crucial conditions are $x < a$ and $y < b$. But the former implies the latter. Since $x < a$ is equivalent to $b' < a$ it is easy to implement a computer search. This tells us that the only pairs (a, b), with $b \leq 50$, for which $b' < a$ are $(9, 16)$, $(11, 30)$, $(16, 22)$, $(29, 37)$. The values of x and y for the first pair are 3 and 1, respectively. Because $xxx_m > 111_{16}$ for $m > 16$, this yields the least solution, which is 111_{16} or 333_9. In base 10, the number is 273.

86. A Highly Divisible Number

An integer that uses each of 0 through 9 once is necessarily divisible by 9 and 3. Divisibility by 2 and 5 implies that the units digit is 0, which takes care of divisibility by 6. Divisibility by 8 (which takes care of 4) means that the two-digit number formed by the hundreds and tens digits must be divisible by 4.

Now, if the solution is 12345∗∗∗∗0, then the aforementioned divisible-by-4 number must be one of 68, 76, 96. This leads to six possibilities—such as 12345976**8**0—none of which is divisible by 7. Our next best hope therefore begins 12346∗∗∗∗0, but this is impossible since none of 58, 78, or 98 is divisible by 4. Moving up to 12347∗∗∗∗0, the possibilities for the hundreds and tens digits are, in order of desirability, 96, 68, and 56. Since 12347589**6**0 is not divisible by 7 while 12347596**8**0 is, the answer is 1234759680. A bonus: this number is divisible by 11 and 12 too.

Problem 86.1. What is the largest number meeting the conditions of the problem?

87. Bilingual Palindromes

There are three solutions: the standard years 232, 353, and 464. To get them, note that the problem splits into two parts, corresponding to December and June, since the Hebrew translation constant is 3761 in December and 3760 in June. We therefore seek a palindromic integer n such that $n + 3761$ or $n + 3760$ is also a palindrome. Consider 3761 first. The one-digit and two-digit palindromes are easily seen to not work as n. If n is aba, a three-digit palindrome, then a must be 2 or 3 (to yield a 3 or a 4 in the sum's rightmost digit). If a is 2, then b must be 3, leading to $232+3761=3993$; if a is 3 then b must be 5, yielding $353+3761=4114$. For the case of 3760, similar reasoning leads to $464+3760=4224$. It is easy to see that there cannot be any 4-digit solutions for n. And there are no 5-or-more-digit solutions, for suppose $ab\ldots ba$ is such. Then the digits in the sum that correspond to the positions of the bs must be different.

Problem 87.1 [Wag3]. Call a positive integer n *palindrome-happy* if there are infinitely many palindromic integers m such that $m+n$ is palindromic too. Find all palindrome-happy integers.

88. A Very Good Year

There are 21 two-digit primes. But we cannot use, after the first digit, a prime whose leading digit is 2, 4, 5, 6, or 8. This leaves us with 11, 13, 17, 19, 31, 37, 71, 73, 79, and 97 for the interior. Let's be optimistic and assume that we can use all ten of these. This means that if we form the directed graph on the vertices 1, 3, 7, 9 with arrows on the 10 edges indicating legal concatenations, then a path that traverses all edges is what we want (see Figure 76).

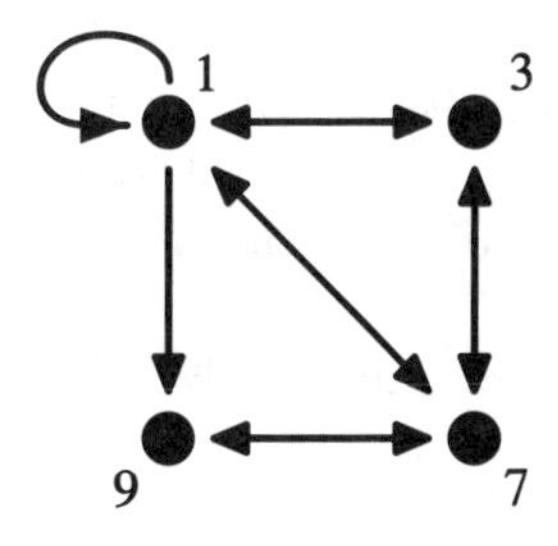

FIGURE 76

Observe that the in-degrees and out-degrees of the four vertices are $(3,4)$, $(2,2)$, $(3,3)$ and $(2,1)$ for 1, 3, 7, and 9, respectively. A path of the sort we seek must begin at a vertex that is more out than in by one, must end at a vertex that is more in than out by one, and can exist only if the other vertices have equal in- and out-degree. Fortunately, this is exactly the situation we are in! Now, if we just start at 1 and greedily move to the largest possible number, we get stuck at 1979. So we switch to 1973, until we get stuck again at 197379. Backtracking again leads us to 197371, but since we must save 9 for the final step, we avoid 9, and also 7, at this stage, leading to 19737131179. Since 61 is the largest prime ending in 1, the answer is 619737131179.

89. Digital Diversity and Unbiased Numbers

There are 9 DD 1-digit numbers and, if n is between 2 and 10, the number of n-digit DD numbers is $9\cdot 9\cdot 8\cdot 7\cdot 6\cdots(11-n)$. This yields the data in the following table, which shows that an unbiased number must exist somewhere between 10,000 and 100,000 (DR stands for digitally repetitive, meaning not DD).

	DD	DR	Total DD	Total DR	Deficit	Top number
1 digit	9	0	9	0	9	9
2 digits	81	9	90	9	81	99
3 digits	648	252	738	261	477	999
4 digits	4536	4464	5274	4725	549	9999
5 digits	27216	62784	32490	67509	–35019	99999

It is easy to see that DD maintains its lead from 10 to 99. Between 100 and 999, in each interval of 100, there are 72 DDs and 28 DRs (72 from $9 \cdot 8$). This can never make up the initial deficit of 81, so the DRs do not catch up below 1000. In each interval of 1000 between 1000 and 9999 there are 504 DDs and 496 DRs, and because there are more DDs than DRs in the first century of each thousand, the deficit of 477 is never erased, and we can be sure that the DD lead is preserved at each step through 9999. Between 10000 and 10999 there are 336 ($8 \cdot 7 \cdot 6$) DDs and 664 DRs, so that the DR deficit is only 221 at 10999. Since all integers in the next thousand are DRs, 11221 is an unbiased number: there are 5610 DDs and 5610 DRs below it.

To prove uniqueness, observe that up to 11,999 the DRs increase their lead, and from 12000 to 99,999 the DDs pick up only 336 in each thousand, so they don't catch up. Things get only worse for the DDs beyond 100,000. Thus 11,221 is the unique unbiased number for base-10 diversity.

Problem 89.1. In base 2, as in base 10, there is a unique unbiased number: 5, because of 1_2, 10_2, 11_2, 100_2. How many unbiased numbers are there in base 8?

90. A Really Big Number

Let us say that integers m and n are equivalent, $m \approx n$, if they have the same rightmost nonzero digit. It is not generally true that if $a \approx b$ then $ac \approx bc$; for example, $2 \approx 12$ but $2 \cdot 5 = 10$ and $12 \cdot 5 = 60$. But if none of the rightmost nonzero digits is 5, then no zeros can be introduced and we get the following useful and easy-to-prove assertion.

Lemma 1. *If 5 is not the rightmost nonzero digit of a, b, or c, then $a \approx b$ implies $ac \approx bc$.*

This leads us to a more subtle fact. Note that there are more factors of 2 than of 5 in $n!$ provided $n > 1$, and so the rightmost nonzero digit of $n!$ is one of 1, 2, 4, 6, or 8.

Lemma 2. $(5n)! \approx 2^n n!$.

Proof.

$$
\begin{aligned}
(5n)! &= 1\cdot 2\cdot 3\cdot 4\cdot \mathbf{5}\cdot 6\cdot 7\cdot 8\cdot 9\cdot \mathbf{10}\cdot 11\cdot 12\cdot 13\cdot 14\cdot \mathbf{15}\cdots(5n-1)\cdot(\mathbf{5}\boldsymbol{n})\\
&= 1\cdot 1\cdot 3\cdot 4\cdot \mathbf{10}\cdot 3\cdot 7\cdot 8\cdot 9\cdot \mathbf{20}\cdot 11\cdot 6\cdot 13\cdot 14\cdot \mathbf{30}\cdots\\
&\qquad \tfrac{1}{2}(5n-4)(5n-3)(5n-2)(5n-1)\cdot(\mathbf{10}\boldsymbol{n})\\
&\approx 1\cdot 1\cdot 3\cdot 4\cdot \mathbf{1}\cdot 3\cdot 7\cdot 8\cdot 9\cdot \mathbf{2}\cdot 11\cdot 6\cdot 13\cdot 14\cdot \mathbf{3}\cdots\\
&\qquad \tfrac{1}{2}(5n-4)(5n-3)(5n-2)(5n-1)\cdot \boldsymbol{n}\\
&= (1\cdot 1\cdot 3\cdot 4)(3\cdot 7\cdot 8\cdot 9)(11\cdot 6\cdot 13\cdot 14)\cdots\\
&\qquad \left(\tfrac{1}{2}(5n-4)(5n-3)(5n-2)(5n-1)\right)\cdot \boldsymbol{n}!\\
&\approx 2\cdot 2\cdot 2\cdots 2\cdot \boldsymbol{n}!\\
&\qquad \text{(by Lemma 1, since the rightmost nonzero digit of } n! \text{ is not 5)}\\
&= 2^n \boldsymbol{n}!
\end{aligned}
$$

Now, eight applications of Lemma 2 reduce $1000000!$ to $2\cdot 2!$ as follows.

$$
\begin{aligned}
1000000! &\approx 2^{200000}\,200000!\\
&\approx 2^{200000}\,2^{40000}\,40000!\\
&\approx 2^{200000}\,2^{40000}\,2^{8000}\,8000!\\
&\approx 2^{200000}\,2^{40000}\,2^{8000}\,2^{1600}\,1600!\\
&\approx 2^{200000}\,2^{40000}\,2^{8000}\,2^{1600}\,2^{320}\,320!\\
&\approx 2^{200000}\,2^{40000}\,2^{8000}\,2^{1600}\,2^{320}\,2^{64}\,64!\\
&= 2^{200000}\,2^{40000}\,2^{8000}\,2^{1600}\,2^{320}\,2^{64}\,60!\cdot 61\cdot 62\cdot 63\cdot 64\\
&\approx 2^{249984}\,2^{60}\,12!\cdot 4\\
&= 2^{250044}\,10!\cdot 11\cdot 12\cdot 4\\
&\approx 6\cdot 2^2\cdot 2!\cdot 8 \quad \text{(because any power of 16 ends in a 6)}\\
&= 24\cdot 8\cdot 2! \approx 2\cdot 2! = 4.
\end{aligned}
$$

Problem 90.1 [WWo]. Let d_n be the rightmost nonzero digit in $n!$. Is the sequence $d_0, d_1, d_2, \ldots$ eventually periodic?

Notes. Problem 90 can be solved by a computer using a fairly straightforward iterative algorithm. In order to eliminate that sort of solution, the problem could just as well be posed for the factorial of a googol (10^{100}). This will defeat any program based on counting from 1 to 10^{100}. But, as observed by Frank Bernhart (Rochester, NY), the ideas of the given solution lead to a more subtle general algorithm that will work even for n as large as a googol.

Arguing exactly as in the given solution, we divide n by 5 to get a quotient q and a remainder r. The problem then reduces to $2^q q! r!$ (consider three cases: q even, q odd and $r \neq 1$, q odd and $r = 1$). Now, we accumulate the 2s as they come up, reducing modulo 4 at each step. Then 2^q is almost right; the exception is when q is 0, for 2^0 gives 1, when we really want 6 since 2^{4k} is a power of 16 and so ends in a 6. This is fixed by substituting 6 for 1 at the appropriate point. The result is then multiplied by `rProd`; the units digit of the result is the desired digit. The 6-substitution hurts us when n is 0 or 1, so these cases are dealt with separately. Here is a *Mathematica* implementation that uses `While`; it could also be done by recursion, but that is a little slower.

```
LastFactorialDigit[0 | 1] = 1

LastFactorialDigit[n_] := Module[{rProd = 1, m = n, q = 0},
  While[m > 0,
    rProd = Mod[rProd * Mod[m, 5]!, 10];
    m = Quotient[m, 5];
    q = Mod[q + m, 4] ];
  Mod[(2^q /. 1 -> 6) * rProd, 10]]

LastFactorialDigit[10^6]
4

LastFactorialDigit[10^100] // Timing
{0.1 Second, 6}
```

The solution presented can be made a little more systematic by looking at different bases (though this is one problem where base two is totally uninteresting). In fact, in the case of a prime base there is a formula, whose derivation we leave as an exercise.

Problem 90.2 (I. Vardi). Show that if p is a prime number then the rightmost nonzero digit when $n!$ is written in base p is given by:

$$(-1)^{a_1+a_3+\cdots+a_{2r-1}} \cdot a_0! \cdot a_1! \cdot a_2! \cdots a_k! \bmod p,$$

where a_i is the coefficient of p^i in the base-p expansion of n, k is the highest power of p that is less than or equal to n, and $2r-1$ is the largest odd number less than or equal to k.

Problem 90.1 was a proposed Olympiad problem, and also appeared in a Russian book on Olympiad problems (see [WWo]). And related work can be found as Exercise 4.40 in [GKP].

8.3 Diophantine Equations

91. The Careless Dealer

Let x be the number of cards that had been dealt so far to South. There are four possibilities to be considered, depending on who received the last card dealt. Suppose first that the last card was dealt to West. Then North, East, and South had received x cards so far, West had received $x+1$ cards, $\frac{2}{3}(x+1)$ cards had been dropped, and South was holding $\frac{3}{2}x$ undealt cards in his hand. Of course the total number of cards must be 52, so $4x+1+\frac{2}{3}(x+1)+\frac{3}{2}x=52$. Solving for x we find that $x=302/37$. But x is an integer, so this case is impossible.

Suppose now that the last card was dealt to North. Then East and South had received x cards so far, West and North had received $x+1$ cards, $\frac{2}{3}(x+1)$ cards had been dropped, and South still had $\frac{3}{2}x$ cards in his hand. Once again the total must be 52, so $4x+2+\frac{2}{3}(x+1)+\frac{3}{2}x=52$. Solving for x we find that $x=8$. Thus, if the last card was dealt to North, then East and South had been dealt 8 cards, West and North had been dealt 9 cards, there were 6 cards on the floor, and South was holding 12 undealt cards. The total number of cards dealt so far was 34. A similar analysis for the remaining two cases shows that neither of them is possible. Therefore the number of cards that had been dealt must be 34.

92. Three Numbers and their Cubes

We first note that $(1,1,1)$ is a solution. In general, one of x, y, z must be positive; assume it is x. And because the only three cubes of nonnegative integers that sum to 3 are 1, 1, 1, we may assume that one of y or z is negative; assume it is y. Because $z=3-(x+y)$, we have $3=x^3+y^3+\left(3-(x+y)\right)^3$, which simplifies to $(x-3)(y-3)(x+y)=8$. Thus $y-3$ divides 8 and y is negative, which means that y is -1 or -5. If $y=-1$, then $(x-3)(x-1)=-2$, which is impossible since x is a positive integer. For $y=-5$, we have $(x-3)(x-5)=-1$, which means $x = 4$. The complete solution set is therefore $(1,1,1)$, $(-5,4,4)$, $(4,-5,4)$, and $(4,4,-5)$.

Notes. In fact, in 1953 L. J. Mordell observed that the four solutions given are the only known integer solutions to $x^3 + y^3 + z^3 = 3$ (see [BK]). Problem 92 is due to A. Brauer and A. Kempner [BK], and also appears as problem 170 in [Sie].

93. A Sum of Fractions

(a) Note that $\frac{x}{y} \cdot \frac{y}{z} \cdot \frac{z}{x} = 1$, so the fractions are not all less than 1. Therefore one of them is greater than or equal to 1 and the sum is therefore greater than 1.

(b) The arithmetic-geometric mean inequality gives

$$\frac{\frac{x}{y} + \frac{y}{z} + \frac{z}{x}}{3} \geq \sqrt[3]{\frac{x}{y} \cdot \frac{y}{z} \cdot \frac{z}{x}} = 1,$$

which implies that the sum of fractions is at least 3. In fact, this sum is 3 if and only if $x = y = z$.

Notes. The parts of this problem appear as problems 153, 154, and 155 in [Sie]. Several people have investigated the question: For which integers n are there integers x, y, z such that $\frac{x}{y} + \frac{y}{z} + \frac{z}{x} = n$? There are solutions when $n = 5$ or 6: $\frac{1}{2} + \frac{2}{4} + \frac{4}{1} = 5$ and $\frac{2}{12} + \frac{12}{9} + \frac{9}{2} = 6$ (J. Browkin; see [Sie, p. 80]). Woody Dudley (DePauw University) has found that $n = 41$ is possible: $\frac{350}{196} + \frac{196}{5} + \frac{5}{350} = 41$. Andrew Bremner and Richard Guy [BG] have gone much farther with this, observing that the question is intimately related to questions of rational points on certain elliptic curves. This has led them to the following list of n for which solutions exist: {3, 5, 6, 9, 10, 13, 14, 15, 16, 17, 18, 19, 20, 21, 26, 29, 30, 31, 35, 36, 38, 40, 41, 44, 47, 51, 53, 54, 57, 62, 63, 64, 66, 67, 69, 70, 71, 72, 73, 74, 76, 77, 83, 84, 86, 87, 92, 94, 96, 98, 99, −4, −9, −10, −11, −12, −16, −17, −21, −22, −24, −25, −27, −28, −29, −32, −33, −34, −35, −36, −37, −38, −40, −44, −45, −46, −47, −48, −49, −50, −53, −55, −56, −57, −59, −60, −63, −64, −65, −66, −67, −68, −72, −73, −76, −77, −79, −80, −81, −82, −84, −85, −86, −87, −88, −89, −90, −92, −94, −95, −100}. For example, $(-1, 10, 100)$ is a solution for $n = -100$, while $(450, 11988, 6845)$ yields a representation for 17. Moreover, their paper gives reasons for their strong belief that this list is complete between -100 and 100.

94. Quadruplets with Square Triplets

Here is a solution found by Diophantus (A.D. 250) (see [Dic, 275 276]). Take any 4 square integers, say a^2, b^2, c^2, d^2, and let n be their sum. Then the 4 rationals $\frac{n}{3} - a^2$, $\frac{n}{3} - b^2$, $\frac{n}{3} - c^2$, $\frac{n}{3} - d^2$ are such that any 3 sum to a square; for example, the first 3 sum to $n - (a^2 + b^2 + c^2)$, which is d^2. To solve the problem we need 4 squares such that the numbers in this solution are positive; that is, we want $\frac{n}{3}$ to be larger than each of the squares. Using square integers, this happens when the four squares are 64, 81, 100, and 121, since their sum is 366 and $366/3 = 122 > 121$. This yields the solution: 1, 22, 41, 58.

Notes. The method of Diophantus generalizes immediately to yield n positive integers such that any $n - 1$ of them sum to a perfect square. The method might yield rationals instead of integers, but scaling up takes care of that. In 1848 C. Gill ([Gil, pp. 60–64], [Dic, p. 456]) found five distinct integers, the sum of every three of which is square. However, his method, which involves a very complicated formula, does not necessarily produce positive integers; a typical example is $\{1917678, 2052219, 4152603, -1981797, 70203\}$. Gill conjectured that his formula would lead to a positive solution. Using *Mathematica* to implement Gill's ideas, and ignoring his warning that "the complexity of the formulas do not encourage the attempt," Wagon [Wag4] found the following five integers near 10^{20}, every three of which sum to a square:

26072 32331 15686 61931
43744 83974 22825 91947
1 18132 65441 36751 38222
1 86378 73280 75870 76747
5 19650 11481 49050 02347

For more background on this sort of problem, see [Guy, problem D15].

Problem 94.1. Show that the method of the solution to Problem 94 produces all examples of 4-tuples with square triplet-sums.

95. When Does The Perimeter Equal The Area?

Let the legs of the triangle have lengths a and b, and the hypotenuse have length c. Since the area and perimeter are equal we have $\frac{1}{2}ab = a+b+c$, so $c = \frac{1}{2}ab - a - b$. The Pythagorean theorem then gives us

$$a^2 + b^2 = (\tfrac{1}{2}ab - a - b)^2 = a^2 + b^2 + 2ab - a^2b - b^2a + \tfrac{1}{4}a^2b^2,$$

or $8ab - 4a^2b - 4b^2a + a^2b^2 = 0$. Dividing by ab yields $(a-4)(b-4) = 8$, so $a - 4$ must divide 8 and therefore the only possible values for a are 2, 3, 5, 6, 8, and 12. Solving for b and c in each case we find that the only other solution is the triangle with side lengths 6, 8, and 10.

Problem 95.1. Show that a triangle has equal numerical values for area and perimeter if and only if the radius of its inscribed circle is 2.

Notes. This problem appeared in [Uma]. For more on equality of perimeter and area, see [Smi].

96. Amicable Rectangles

Suppose $a \times b$ and $c \times d$ define an amicable pair, with $a \le b$, $c \le d$, and $a \le c$. Then $2(a+b) = cd$ and $2(c+d) = ab$. The second equation yields $d = \frac{1}{2}ab - c$, which we can the substitute into the first to get:

$$c^2 - \frac{ab}{2}c + 2(a+b) = 0, \tag{1}$$

which yields $b = (2c^2 + 4a)/(ac - 4)$; there is a similar expression for d in terms of a and c. Now, we prepare for a computer search by finding bounds on the possible values for a and c. First, we complete the square in (1) to get:

$$(c - ab/4)^2 = a^2b^2/16 - 2(a+b), \quad \text{or} \quad (ab - 4c)^2 = a^2b^2 - 32(a+b). \tag{2}$$

Because $c \le d$, we have $ab - 4c = 2d - 2c \ge 0$; also $c \ge 1$, so $0 \le ab - 4c \le ab - 4$. Thus (2) implies $(ab-4)^2 \ge a^2b^2 - 32(a+b)$, which reduces to $4a + 2 \ge b(a-4)$. But $a \le b$, so $4a + 2 \ge a(a-4)$, or $a^2 - 8a - 2 \le 0$, or $(a-4)^2 \le 18$, or $a \le 8$. By symmetry, $c \le 8$ as well, so we have $1 \le a \le c \le 8$.

We can go farther (more precisely, less far) by appealing now to $a \le c$. For then we have $ab - 4a \ge ab - 4c = 2(d - c) \ge 0$, and so, from (2), $(ab - 4a)^2 \ge a^2b^2 - 32(a+b)$, which reduces to $2a \ge b(a-2)$. But $a \le b$, so $2a \ge a(a-2)$, or $a^2 - 4a \le 0$, or $a \le 4$.

So we now know that (a, c) must satisfy $1 \le a \le 4$ and $a \le c \le 8$; there are only 26 such pairs and a quick computer check shows that 7 of them lead to integer values of b and d. The complete list of amicable pairs of rectangles is therefore:

1×34 and 7×10	1×38 and 6×13	1×54 and 5×22
2×10 and 4×6	2×13 and 3×10	3×6 and 3×6
4×4 and 4×4		

Note. This problem is due to Kenneth Wilke [Wil1].

97. Strange Boxes

Suppose integers x, y, z are the dimensions of a box and assume $x \geq y \geq z$. Then the problem requires that

$$2(xy + xz + yz) = xyz. \tag{1}$$

This means $xyz \leq 6xy$ so $z \leq 6$. Also, $(z-2)xy = 2xz + 2yz$, which is positive, so $z \geq 3$. Thus $z \in \{3, 4, 5, 6\}$.

Now, (1) yields

$$xy = \frac{2z}{z-2}(x+y) \leq \frac{4z}{z-2}x,$$

so $y \leq \dfrac{4z}{z-2}$. Moreover, solving (1) for x yields $x = \dfrac{2yz}{yz - 2y - 2z}$, which must be a positive integer, so the denominator is positive and $y > \dfrac{2z}{z-2}$. Thus y lies in

$$\left(\frac{2z}{z-2}, \frac{4z}{z-2}\right].$$

If $z = 3$ the choices for y are 7, 8, 9, 10, 11, 12. Taking $y = 11$ would imply a noninteger value for x, since $6 \cdot 11$ is not divisible by 5. But the other choices for y all work, yielding the boxes: $(42, 7, 3)$, $(24, 8, 3)$, $(18, 9, 3)$, $(15, 10, 3)$, and $(12, 12, 3)$. If $z = 4$, y must be one of 5, 6, 7, or 8 and only 5, 6, and 8 work, yielding the boxes $(20, 5, 4)$, $(12, 6, 4)$, and $(18, 8, 4)$. Similarly, $z = 5$ leads to $(10, 5, 5)$ and $z = 6$ to $(6, 6, 6)$, for a total of 10 strange boxes. If symmetry is ignored, there are 49 solutions.

Notes. The volumes of the solutions are all distinct, except for the pair $(24, 8, 3)$ and $(18, 8, 4)$, which each have volume and surface area equal to 576. Note also that the problem is equivalent to asking for all solutions of $\frac{1}{x} + \frac{1}{y} + \frac{1}{z} + \frac{1}{w} = 1$ with $w = 2$. As such, it appears as problem 160 in [Sie]; see also Problem 103 in the present book.

98. Dividing a Product by a Sum

The answer is all N except those of the form $p-1$, where p is a prime larger than 2. To see why, note first that $1+2+3+\cdots+N = N(N+1)/2$. Thus, we must find all positive integers N for which $N!$ is divisible by $N(N+1)/2$, or equivalently $2(N-1)!$ is divisible by $N+1$. Clearly if $N+1$ is a prime larger than 2 then $2(N-1)!$ is not divisible by $N+1$, and if $N+1=2$ then also $2(N-1)!=2$, so $2(N-1)!$ is divisible by $N+1$. Now suppose $N+1$ is not prime. Then $N+1$ can be written as a product $N+1=ab$, where $1<a$, $b<N$. If $a\neq b$, then a and b are both among the terms in $2(N-1)!$, so $2(N-1)!$ is divisible by ab, which equals $N+1$. Now suppose $a=b$, so $N+1=a^2$. If $a>2$ then $2a<N$ and both a and $2a$ are among the terms in $2(N-1)!$, so $2(N-1)!$ is divisible by $a^2=N$. The only case not yet covered is when $a=2$. In this case $N=3$ and $2(N-1)!=4=N+1$, so $2(N-1)!$ is divisible by $N+1$.

99. They're in the Money

Let a_1 be the number of pennies Alice got, a_2 the number of her half-dollars, and a_3 the number of her silver dollars; define b_i for Bob similarly and let $d_i=a_i-b_i$. Then each d_i is an integer and $d_1+d_2+d_3=0=d_1+50d_2+100d_3$. It follows that 50 divides d_1 and, by subtraction, $49d_2+99d_3=0$. This means that 49 divides d_3 and 99 divides d_2. Thus there are integers n_i such that $d_1=50n_1$, $d_2=99n_2$, and $d_3=49n_3$.

Now, because $50n_1+99n_2+49n_3=0=n_1+99n_2+98n_3$, we have, by subtraction, $49n_1=49n_3$, whence $n_1=n_3$. This means that $99n_1+99n_2=0$, whence $n_2=-n_1$. Thus we may conclude that $n_2\neq 0$, so that Alice and Bob received different numbers of half-dollars.

Assume that Alice received more half-dollars than Bob (otherwise reverse their roles). Then n_2 is positive and so, because $b_2=a_2-99n_2$, a_2 must be at least as large as 99. Trying $a_2=99$ yields the solution in which Alice gets 99 half-dollars and nothing else while Bob receives 50 pennies and 49 silver dollars. Since $a_2\geq 99$, this solution is minimal and we can be sure that each check was worth at least \$49.50.

100. Lucky Numbers

A key observation is that if n is lucky then so is $2n+2$. For if $n=a+b+\cdots$ is a lucky decomposition, then $\frac{1}{2}+\frac{1}{2}(\frac{1}{a}+\frac{1}{b}+\cdots)=1$ and so $2+2a+2b+\cdots=2+2n$ is a lucky decomposition. Similarly, $3+6+2a+2b+\cdots$ is a lucky decomposition of $2n+9$. Now, if we knew that all integers between 24 and 55 are lucky, we could use these two formulas to conclude that $50, 57, 52, 59, 54, 61, \ldots$ ad infinitum are all lucky. Thus it remains only to show that all integers between 24 and 55 are lucky, and to determine which integers less than 24 are unlucky.

Therefore $(p+y)q = r(p+x)$, so necessarily q divides $p+x$ and r divides $p+y$. By symmetry, p divides $q+x$ and p divides $r+y$.

The equations $pr = yq+1$ and $p < q$ tell us that $yq < pr < rq$, so $y < r$, and hence $p+y < 2r$. Since r divides $p+y$, r must in fact equal $p+y$. By (1), $p^2 - xy = 1$ so $q = p+x$; since p divides $q+x$, this means p divides $p+2x$, so p divides $2x$. Since $pq - xr = 1$, p and x are relatively prime, and hence p must divide 2, so p is 1 or 2. But $p = 1$ implies $xy = 0$ (since $p^2 - xy$ must be positive), which is a contradiction for if x is 0, then $q = 1$, and if $y = 0$ then $r = 1$. Therefore $p = 2$, whence $4 - xy = 1$ implies $xy = 3$. Since $x < y$ this means $x = 1$ and $y = 3$, so $q = p+x = 3$ and $r = p+y = 2+3 = 5$.

107. An Odd Set of Positive Integers

Let S consist of all positive integers except $\{1, 2, 4, 7, 10\}$;

$$S = \{3, 5, 6, 8, 9, 11, 12, 13, 14, \ldots\}.$$

Then each of 3, 5, 6, 8, 9, 11, 12 is a sum of two numbers not in S; $13 = 8+5$, and any integer n greater than 13 has the form $3 + (n-3)$. Conversely, none of 1, 2, 4, 7, or 10 is a sum of numbers in S or numbers not in S. Therefore this set solves the problem. It is easy to see that the solution is unique, since 1 cannot be in S, 2 cannot be in S, 3 must be in S, and so on.

108. Sums of Squares

The largest integer that cannot be written as a sum of two or more distinct positive squares is 128. To see why, we first prove a lemma.

Lemma. *Suppose n is a positive integer, and every integer from $n+1$ to $4n+35$ (inclusive) can be written as a sum of distinct positive squares. Then every integer larger than n can be written as a sum of distinct positive squares.*

Proof. Suppose not, and let k be the least counterexample. By hypothesis, $k \geq 4n+36$. Write k in the form $k = 4q+r$, with $0 \leq r \leq 3$, and note that $n+9 \leq q < k$. We now consider the four possible values for r.

Suppose first that $r = 0$. By the minimality of k, q can be written as a sum of distinct positive squares; say $q = a_1^2 + a_2^2 + \cdots + a_m^2$. But then $k = 4q = (2a_1)^2 + (2a_2)^2 + \cdots + (2a_m)^2$, contradicting the assumption that k cannot be written as a sum of distinct positive squares. If $r = 1$ then similar reasoning leads to $k = 4q + 1 = (2a_1)^2 + (2a_2)^2 + \cdots + (2a_m)^2 + 1$, which is again a sum of distinct positive squares.

If $r = 2$ then $k = 4q + 2 = 4(q - 2) + 10$. Since $n + 7 \le q - 2 < k$ and k was chosen to be minimal, $q - 2$ can be written as a sum of distinct positive squares; say $q - 2 = b_1^2 + b_2^2 + \cdots + b_m^2$. Then

$$k = 4(q - 2) + 10 = (2b_1)^2 + (2b_2)^2 + \cdots + (2b_m)^2 + 1 + 9,$$

which is another sum of distinct positive squares.

Finally, if $r = 3$ then $k = 4q + 3 = 4(q - 8) + 1 + 9 + 25$, and similar reasoning shows that this is a sum of distinct positive squares. This completes the proof of the lemma.

By the lemma, we can find the largest integer which is not a sum of distinct positive squares by searching for an n such that n is not a sum of distinct positive squares, but all integers from $n + 1$ to $4n + 35$ are. The following *Mathematica* code implements this search, and produces a list of all integers that are not a sum of distinct positive squares. The function `SumSquares[n]` produces a list of all integers that can be written as a sum of squares of at least two distinct positive integers less than or equal to n. `SumSquaresQ[n]` is true if and only if n can be written as a sum of two or more distinct positive squares. The list of integers that cannot be written as a sum of distinct positive integers is produced in the variable `NonSumSquares`. The phrase "`SumSquares[n] =`" is used to cache values as they are computed, thus avoiding a major slowdown of the computation because of recomputation of values.

```
SumSquares[1] = {}
SumSquares[n_] := (SumSquares[n] =
 Union[SumSquares[n-1], SumSquares[n-1] + n^2, n^2 + Range[n-1]^2])
SumSquaresQ[n_] := MemberQ[SumSquares[Floor[Sqrt[n-1]]], n]

a = 1;
NonSumSquares = {1};
While[a < 4 * Last[NonSumSquares] + 35,
  a = a + 1;
  If[!SumSquaresQ[a], AppendTo[NonSumSquares, a]]]
```

After this code has been executed, `NonSumSquares` contains the list {1, 2, 3, 4, 6, 7, 8, 9, 11, 12, 15, 16, 18, 19, 22, 23, 24, 27, 28, 31, 32, 33, 36, 43, 44, 47, 48, 60, 64, 67, 72, 76, 92, 96, 108, 112, 128}.

Notes. This problem appeared in both [Sil1] (with a solution different from the one given here) and [Sil2]. More generally, R. Sprague [Spr] showed that for every $n \ge 2$ there is a largest integer that is not expressible as a sum of distinct nth powers. For more on this problem see [Mak, Tri2, Nel].

109. Sums and Differences

The assertion is false. The smallest counterexample example known, due to Imre Ruzsa, is: $\{1,3,4,5,8,12,13,15\}$, for which there are 26 sums (the interval $[2,30]$ excluding 3, 22, and 29), but only 25 differences ($[-14,14]$, excluding ± 6 and ± 13).

Notes. It had been conjectured by John H. Conway that D is never less than S. A 9-element counterexample was found by John Marica [Mar]. However, Sherman Stein [Ste] has pointed out that the existence of counterexamples follows from a "negligible modification" of Sophie Piccard's work [Pic] on measure-theoretic aspects of sums and difference sets.

For Further Investigation. Is there an example having 7 elements? It is known that there are no examples with 6 elements [Ruz].

110. A Radical Equation

$$\begin{aligned}\left(\frac{1-\sqrt{2}+\sqrt{3}}{1+\sqrt{2}-\sqrt{3}}\right)\left(\frac{1+\sqrt{2}+\sqrt{3}}{1+\sqrt{2}+\sqrt{3}}\right) &= \frac{(1+\sqrt{3})^2-2}{(1+\sqrt{2})^2-3} \\ &= \frac{2+2\sqrt{3}}{2\sqrt{2}} \\ &= \frac{1+\sqrt{3}}{\sqrt{2}} \\ &= \frac{\sqrt{2}+\sqrt{6}}{2}.\end{aligned}$$

Thus (x, y) will give a solution if and only if $\sqrt{2}+\sqrt{6} = \sqrt{x}+\sqrt{y}$. Two obvious solutions are $(2,6)$ and $(6,2)$. These are the only solutions. For squaring both sides of $\sqrt{2}+\sqrt{6} = \sqrt{x}+\sqrt{y}$ gives $8+2\sqrt{12} = x+y+2\sqrt{xy}$, so $x+y-8 = 2(\sqrt{12}-\sqrt{xy})$. Because the right side is rational, so is its square, $4(12+xy-4\sqrt{3xy})$, and therefore so is $\sqrt{3xy}$. But this means that $3xy$ is a perfect square, whence $xy = 3n^2$ and $\sqrt{12}-\sqrt{xy} = 2\sqrt{3}-\sqrt{3}n = (2-n)\sqrt{3}$. This can be rational only if $n = 2$, which means $xy = 12$. So, $x+y-8 = 2(\sqrt{12}-\sqrt{12}) = 0$, whence $x+y = 8$. Finally, $xy = 12$ now becomes $x(8-x) = 12$, which proves that $(2,6)$ and $(6,2)$ are the only solutions.

CHAPTER 9

Algebra

9.1 Polynomials

111. An Elusive Quadratic

We may as well try for a polynomial that is symmetric about 2.5, for then we need not worry about $p(3)$ and $p(4)$. Suppose $p(2) = p(3) = n^2$. Then $p(x)$ has the form $p(x) = k(x-2)(x-3)+n^2$, so $p(1) = p(4) = 2k+n^2$ and $p(5) = 6k+n^2$. Therefore we seek n and k such that $2k+n^2$ is a square but $6k+n^2$ is not. Since $2k$ is even and a difference of two squares, we can try $2k = 4-0$, for $k = 2$ and $n = 0$. Then $6k+n^2 = 12$. So $2(x-2)(x-3)$, or $2x^2 - 10x + 12$ works.

112. A Polynomial Fitting Problem

No, there is no such polynomial. In fact, there is not even a polynomial $q(x)$ with integer coefficients such that $q(1) = 2$ and $q(3) = 5$. To see why, suppose $q(x) = a_0 + a_1x + a_2x^2 + \cdots + a_nx^n$ is such a polynomial. Then

$$3 = 5 - 2 = q(3) - q(1) = (3-1)a_1 + (3^2-1)a_2 + \cdots + (3^n-1)a_n.$$

But this is impossible, since for every positive integer k, $3^k - 1$ is even, and therefore the right side of the equation above is even.

113. A Polynomial Oddity

There cannot be such an integer k. For suppose k did exist. Let $Q(x) = P(x) - 2$; then Q is zero at a, b, c, and d, so $Q(x) = (x-a)(x-b)(x-c)(x-d)R(x)$, where $R(x)$ is a polynomial with integer coefficients. (To see that the coefficients of $R(x)$ are all integers, think about computing $R(x)$ by dividing $Q(x)$ by $(x-a)(x-b)(x-c)(x-d)$.) But then $Q(k) = (k-a)(k-b)(k-c)(k-d)L$, which is impossible, since a product of 4 or more distinct integers, positive or negative, cannot be one of -1, 1, 3, 5, or 7 (for example: 7 is 7 or $7 \cdot 1$ or $(-7) \cdot 1 \cdot (-1)$, but is not a product of four distinct integers). Since 9 is such a product, one can get $P(k)$ to be 11: just consider $(x-3)(x+3)(x-1)(x+1)+2$, which is 2 at ± 3, ± 1, and 11 at 0.

Note. A problem quite similar to this appeared in [SCY, problem 212]. It was also used on the 1970 Canadian Olympiad.

114. A Polynomial Pattern

Since $P(k) = \frac{1}{k}$ for $k = 1, 2, \ldots, 9$, each of these nine integers is a root of $xP(x) - 1$, a polynomial of degree 9. Hence

$$xP(x) - 1 = c(x-1)(x-2)(x-3)(x-4)(x-5)(x-6)(x-7)(x-8)(x-9)$$

for some constant c. But equating constant terms yields $-1 = -c \cdot 9!$, whence $c = \frac{1}{9!}$ and $10P(10) - 1 = \frac{9!}{9!}$, or 1. Therefore $P(10) = 1/5$.

Problem 114.1. Suppose $P(k)$ is a polynomial of degree $n-1$ with real coefficients and $P(k) = \frac{1}{k}$ for $k = 1, 2, 3, \ldots, n$. What is $P(n+1)$?

Problem 114.2. Suppose $P(x)$ is a polynomial of degree n such that $P(k) = 2^k$ for $k = 0, 1, \ldots, n$. What is $P(n+1)$?

Note. A similar problem appeared on thc 1975 U.S. Mathematical Olympiad [Kla, pp. 4, 20–21].

115. Powerful Patterns

We start with the formula

$$(a+b+c)^4 = a^4 + b^4 + c^4 + 4(a^3b + a^3c + b^3a + b^3c + c^3a + c^3b) + \\ 6(a^2b^2 + a^2c^2 + b^2c^2) + 12(a^2bc + b^2ac + c^2ab). \qquad (1)$$

The quantity we seek, $a^4 + b^4 + c^4$, appears on the right side, but to find it we must somehow get rid of the unwanted terms. We begin by eliminating $a^2bc + b^2ac + c^2ab$.

This quantity also appears in

$$(a+b+c)^2(a^2+b^2+c^2) = a^4+b^4+c^4+2(a^3b+a^3c+b^3a+b^3c+c^3a+ \\ c^3b+a^2b^2+a^2c^2+b^2c^2+a^2bc+b^2ac+c^2ab). \quad (2)$$

Subtracting 6 times equation (2) from equation (1) yields

$$(a+b+c)^4 - 6(a+b+c)^2(a^2+b^2+c^2) = -5(a^4+b^4+c^4) - \\ 8(a^3b+a^3c+b^3a+b^3c+c^3a+c^3b) - 6(a^2b^2+a^2c^2+b^2c^2). \quad (3)$$

Similarly, we can eliminate the other unwanted terms by adding multiples of $(a^2+b^2+c^2)^2$ and $(a+b+c)(a^3+b^3+c^3)$ to equation (3). This leads us to the equation

$$(a+b+c)^4 - 6(a+b+c)^2(a^2+b^2+c^2) + 3(a^2+b^2+c^2)^2 + \\ 8(a+b+c)(a^3+b^3+c^3) = 6(a^4+b^4+c^4)$$

Thus,

$$\begin{aligned} a^4+b^4+c^4 &= \tfrac{1}{6}\big[(a+b+c)^4 - 6(a+b+c)^2(a^2+b^2+c^2) + \\ &\qquad 3(a^2+b^2+c^2)^2 + 8(a+b+c)(a^3+b^3+c^3)\big] \\ &= \tfrac{1}{6}[3^4 - 6\cdot 3^2\cdot 5 + 3\cdot 5^2 + 8\cdot 3\cdot 7] \\ &= 9. \end{aligned} \quad (4)$$

Notes. This problem appeared in [Wils]; similar problems can be found in [Ahl, Larl]. In fact, this sort of problem has a long history; the 1890 text [HK, p. 442] contains a similar problem. A computer solution of Problem 115 is possible. Consider the following *Mathematica* command:

```
Solve[{w == a + b + c, x == a^2 + b^2 + c^2,
      y == a^3 + b^3 + c^3, z == a^4 + b^4 + c^4}, z, {a, b, c}]
```

This instructs *Mathematica* to solve the given equations for z, eliminating a, b, and c. The result is $z = (w^4 - 6w^2x + 3x^2 + 8wy)/6$, which is the same as the first line of equation (4).

Problem 115.1. Does the pattern continue? Does $a^5+b^5+c^5 = 11$?

Problem 115.2. Show that there are no solutions for a, b, and c in the real numbers, but there are solutions in the complex numbers.

9.2 Miscellaneous

116. Christmas Time

Let ρ and θ be the angles, in degrees, measured clockwise from the vertical position, made by the hour and minute hands, respectively. Then, if time t is measured in minutes since noon, $\rho = t/2$ and $\theta = 6t$. The two hands coincide if and only if, for some integer n, $\theta = \rho + 360n$. This means $6t = t/2 + 360n$, or $t = 720n/11$.

Let t_0 be Santa's start time and t_1 the time he reached 8 miles. Then t_0 is an integer because of the initial position of the second hand and $t_1 = t_0 + \frac{8}{33}60 = t_0 + \frac{160}{11}$ minutes. Since $t_1 = 720n/11$, this gives $t_0 = (720n - 160)/11$ minutes. But 10 is the only integer n for which $720n - 160$ is positive and divisible by 11 and $t_0 < 720$ (the number of minutes in 12 hours); the easiest way to get this is to note that the divisibility condition yields $n \equiv 10 \pmod{11}$. Thus Santa started his trip at $(7200 - 160)/11$, or 640, minutes after noon. This corresponds to 10:40 P.M. He better hurry!

117. Play Ball!

The batter now has 19 hits. Let h and n be the number of hits and number of at-bats, respectively, after the single. Then

$$\frac{h-1}{n-1} + \frac{1}{100} = \frac{h}{n},$$

which yields $100h = n(101 - n)$. Let $m = 101 - n$; then $100h = mn$. Because $n \geq 2$ and $h \geq 2$, both n and m are between 2 and 99, inclusive. Moreover, m and n have opposite parity, so 4 must divide one of them; similarly, if n is divisible by 5 then m isn't, and vice versa, so one of them is divisible by 25. Because m and n are less than 100, one of them is divisible by 4 and the other by 25. We may assume that n is divisible by 25 and m by 4 (otherwise just switch their roles). Since neither $101 - 50$ nor $101 - 75$ is divisible by 4, n must be 25 and m must be $101 - 25 = 76$. It follows that $h = 25 \cdot 76/100$, or 19. Note that in this case the batting average went from .750 to .760. If, as might have happened, $n = 76$ and $m = 25$, then the average went from .240 to .250.

118. The Potato Peelers

Let p be the number of potatoes in each pile initially, a_1 the number in Alice's pile when the ratio is $2:1$ and a_2 the number when the ratio is $7:3$; define b_1 and b_2 similarly for Bob. Now, every two minutes, Alice's pile is reduced by 3 potatoes while Bob's is reduced by 1. Assume first that the $2:1$ ratio is reached in $2n$ minutes, an even number. Then, 5 minutes after the $2:1$ ratio, $a_2 = a_1 - 7$ and $b_2 = b_1 - 3 = 2a_1 - 3$. At that time $7(a_1 - 7) = 3(2a_1 - 3)$, so $a_1 = 40$ and $b_1 = 80$. Moreover, $2(p - 3n) = p - n = 80$ so that $p = 5n$ and $4n = 80$. Hence $n = 20$, it took 40 minutes to get from $1:1$ to $2:1$, and they each started with 100 potatoes.

Now, if it took $2m$ minutes to get from $2:1$ to $3:1$, $3(40 - 3m) = 80 - m$, so $m = 5$, yielding 10 minutes for this portion. Hence the total elapsed time is 50 minutes, at which time Alice has 25 unpeeled potatoes and Bob 75. If instead the $2:1$ to $3:1$ interval was $2m + 1$ minutes long, then $3(40 - 3m - 1) = 80 - m - 1$, or $38 = 8m$, contradiction.

It remains to deal with the case that the $2:1$ ratio is reached after an odd number of minutes, say $2n + 1$. Then 5 minutes after the $2:1$ ratio, $a_2 = a_1 - 8$ and $b_2 = b_1 - 2 = 2a_1 - 2$. At that time $7(a_1 - 8) = 3(2a_1 - 2)$, so $a_1 = 50$ and $b_1 = 100$. Moreover, $2(p - 3n - 1) = p - n - 1 = 100$ so that $p = 5n + 1$ and $4n = 100$. Hence $n = 25$, it took 51 minutes to get to $2:1$, and they each started with 126 potatoes. If it took $2m$ additional minutes to get from $2:1$ to $3:1$, $3(50 - 3m) = 100 - m$, so $50 = 8m$, contradiction. If it took $2m + 1$ additional minutes to get from $2:1$ to $3:1$, $3(50 - 3m - 2) = 100 - m$, so $44 = 8m$, again a contradiction. Thus in this case the $3:1$ ratio cannot arise. So, strictly speaking, there are two possible answers: they started with 100 potatoes and it took 50 minutes for their piles to be in a $3:1$ ratio, or they started with 126 potatoes and *never* reached the $3:1$ ratio.

119. Playing the Stock Market

Alice's average cost per share is always less than or equal to Bob's.

Suppose Alice spends x dollars each month, Bob buys y shares each month, and the price of a share in month i is p_i. Then Alice spends a total of nx dollars to purchase $x/p_1 + x/p_2 + \cdots + x/p_n$ shares, so her average cost per share is

$$\frac{nx}{x/p_1 + x/p_2 + \cdots + x/p_n} = \frac{n}{1/p_1 + 1/p_2 + \cdots + 1/p_n}.$$

In other words, Alice's average cost per share is the harmonic mean of the prices of the stock in the n months.

Meanwhile, Bob spends $y(p_1+p_2+\cdots+p_n)$ dollars to purchase ny shares, so his average cost per share is

$$\frac{y(p_1+p_2+\cdots+p_n)}{ny}=\frac{p_1+p_2+\cdots+p_n}{n}.$$

In other words, Bob's average price per share is the arithmetic mean of the prices of the stock in the n months.

The conclusion now follows from the following lemma.

Lemma. The harmonic mean of a list of positive numbers is always less than or equal to their arithmetic mean. In other words, for any positive numbers $p_1, p_2, \ldots, p_n$,

$$\frac{n}{1/p_1+1/p_2+\cdots+1/p_n}\leq\frac{p_1+p_2+\cdots+p_n}{n}.$$

Proof. It suffices to show that

$$(p_1+p_2+\cdots+p_n)(1/p_1+1/p_2+\cdots+1/p_n)\geq n^2.$$

The left side of this inequality is

$$\sum_{i=1}^{n}p_i\sum_{j=1}^{n}\frac{1}{p_j}=\sum_{i=1}^{n}\sum_{j=1}^{n}\frac{p_i}{p_j}=\sum_{i=1}^{n}\sum_{j=1}^{i-1}\left(\frac{p_i}{p_j}+\frac{p_j}{p_i}\right)+\sum_{i=1}^{n}\frac{p_i}{p_i}.$$

Now, $x+\frac{1}{x}=2+(x-1)^2/x\geq 2$ for positive x (alternatively, use calculus). Thus

$$\sum_{i=1}^{n}p_i\sum_{j=1}^{n}\frac{1}{p_j}\geq\sum_{i=1}^{n}\sum_{j=1}^{i-1}2+\sum_{i=1}^{n}1=\frac{2(n^2-n)}{2}+n=n^2,$$

as required.

Note. The method of using a constant amount of money to buy stocks is called "dollar-cost averaging." As pointed out by the Beardstown Ladies' Investment Club [BLIC, pp. 133–134], "'Dollar cost averaging', popular with conservative, long-term investors, is a stock-purchasing technique to make sure you pay Wal-Mart prices. Dollar cost averaging requires making regular purchases of a particular stock or set of stocks with a set sum regardless of the market's level. ... Over an extended period of time during which stocks have moved both up and down, you will find that your average price per share is lower than the average price for that period.

The mathematical reason: Simply because you bought more shares when prices were low and fewer shares when prices were high."

120. An Impossible Inequality

Let x, y, and z be any real numbers. We consider three cases.

Case 1. $xy + yz + xz = 0$. Then the inequality in the problem is clearly not satisfied, since the fraction is undefined.

Case 2. $xy+yz+xz > 0$. Clearly $(x-y)^2+(y-z)^2+(x-z)^2 \geq 0$. Expanding gives us $2x^2+2y^2+2z^2-2xy-2yz-2xz \geq 0$, so $x^2+y^2+z^2 \geq xy+yz+xz$. Dividing through by $xy + yz + xz$ yields

$$\frac{x^2+y^2+z^2}{xy+yz+xz} \geq 1,$$

contradicting the second half of the proposed inequality.

Case 3. $xy+yz+xz < 0$. In this case we start with $(x+y+z)^2 \geq 0$. Expanding, we get $x^2+y^2+z^2+2xy+2yz+2xz \geq 0$, so $x^2+y^2+z^2 \geq -2(xy+yz+xz)$. Dividing through by $xy + yz + xz$ and reversing the direction of the inequality (since $xy + yz + xz < 0$) we get

$$\frac{x^2+y^2+z^2}{xy+yz+xz} \leq -2,$$

which contradicts the first half of the proposed inequality.

121. Radical Multiples

First note that for any positive integer n, the integers P and Q are uniquely determined by the equation $(1+\sqrt{2})^n = P + Q\sqrt{2}$. To see why, suppose that $(1+\sqrt{2})^n = P+Q\sqrt{2} = P'+Q'\sqrt{2}$. Then $P-P' = (Q'-Q)\sqrt{2}$, so if $Q' \neq Q$ then $(P-P')/(Q'-Q) = \sqrt{2}$, contradicting the irrationality of $\sqrt{2}$. Thus $Q' = Q$, and it follows that $P' = P$.

We can now find formulas for P and Q by multiplying out $(1+\sqrt{2})^n$, using the binomial theorem:

$$\begin{aligned}(1+\sqrt{2})^n &= \binom{n}{0} + \binom{n}{1}\sqrt{2} + \binom{n}{2}(\sqrt{2})^2 + \binom{n}{3}(\sqrt{2})^3 + \cdots + \binom{n}{n}(\sqrt{2})^n \\ &= \left[\binom{n}{0} + 2\binom{n}{2} + 4\binom{n}{4} + \cdots\right] + \left[\binom{n}{1} + 2\binom{n}{3} + 4\binom{n}{5} + \cdots\right]\sqrt{2}.\end{aligned}$$

By the uniqueness of P and Q it follows that

$$P = \left[\binom{n}{0} + 2\binom{n}{2} + 4\binom{n}{4} + \cdots\right] \quad \text{and} \quad Q = \left[\binom{n}{1} + 2\binom{n}{3} + 4\binom{n}{5} + \cdots\right].$$

The key to solving the problem is to carry out a similar expansion for $(1-\sqrt{2})^n$:

$$\begin{aligned}(1-\sqrt{2})^n &= \binom{n}{0} - \binom{n}{1}\sqrt{2} + \binom{n}{2}(\sqrt{2})^2 - \binom{n}{3}(\sqrt{2})^3 + \cdots + \binom{n}{n}(-\sqrt{2})^n \\ &= \left[\binom{n}{0} + 2\binom{n}{2} + 4\binom{n}{4} + \cdots\right] - \left[\binom{n}{1} + 2\binom{n}{3} + 4\binom{n}{5} + \cdots\right]\sqrt{2} \\ &= P - Q\sqrt{2}.\end{aligned}$$

But $\sqrt{2} - 1 < \frac{1}{2}$, so $|P - Q\sqrt{2}| = (\sqrt{2}-1)^n < \frac{1}{2}$. Thus, P is an integer that differs from Q$\sqrt{2}$ by less than $\frac{1}{2}$, so it must be the integer nearest to $Q\sqrt{2}$.

Note. A similar problem appears in [SCY, problem 104].

Problem 121.1. Since P is close to $Q\sqrt{2}$, it is reasonable to expect that P/Q should be a good approximation to $\sqrt{2}$. Show that P/Q differs from $\sqrt{2}$ by less than $1/Q^2$.

122. Pascal's Determinant

We proceed by induction on n. The result clearly holds when $n = 1$. Now suppose it holds for some value of n, and consider the matrix consisting of the first $n+1$ rows and columns of the array.

First we subtract from each row except the first the row before it, which leaves the determinant unchanged and gives us:

$$\begin{matrix} 1 & 1 & 1 & 1 & 1 & \cdots \\ 0 & 1 & 2 & 3 & 4 & \cdots \\ 0 & 1 & 3 & 6 & 10 & \cdots \\ 0 & 1 & 4 & 10 & 20 & \cdots \\ 0 & 1 & 5 & 15 & 35 & \cdots \\ \vdots & & & & & \end{matrix}$$

Now we subtract from each column except the first the column before it:

$$\begin{matrix} 1 & 0 & 0 & 0 & 0 & \cdots \\ 0 & 1 & 1 & 1 & 1 & \cdots \\ 0 & 1 & 2 & 3 & 4 & \cdots \\ 0 & 1 & 3 & 6 & 10 & \cdots \\ 0 & 1 & 4 & 10 & 20 & \cdots \\ \vdots & & & & & \end{matrix}$$

An expansion by minors of the first row now yields the desired result, since the inductive hypothesis applies to the $n \times n$ matrix obtained by crossing out the first row and column.

123. A Shuffling Reconstruction

The device can be thought of as a permutation σ that takes the card in position n and moves it to position $\sigma(n)$. Then σ^2 is the cycle

$$(1 \to 8 \to 4 \to 7 \to 13 \to 5 \to 9 \to 2 \to 12 \to 3 \to 6 \to 11 \to 10 \to 1).$$

It follows that σ is also a 13-cycle, for if σ had any smaller cycles, they would show up in σ^2. But then σ^{13} is the identity, so

$$\begin{aligned} \sigma &= (\sigma^2)^7 \\ &= (1 \to 2 \to 8 \to 12 \to 4 \to 3 \to 7 \to 6 \to 13 \to 11 \to 5 \to 10 \to 9 \to 1). \end{aligned}$$

Thus the order of the cards after the first shuffle was:

9, A, 4, Q, J, 7, 3, 2, 10, 5, K, 8, 6.

Problem 123.1 (H. Dinh). Using the same setup as in Problem 123, suppose that after a number of shuffles, the order of the cards from top to bottom is

5, 9, Q, 8, K, 3, 4, A, 10, J, 6, 2, 7.

At least two and no more than ten shuffles have been made, but the exact number is unknown. What was the order of the cards after the first shuffle?

124. Associativity and Commutativity

First, note that since A commutes with both B and C, in any $*$-product of As, Bs, and Cs, the As can be moved anywhere without changing the value of the product. We will call this the *A-movement* rule. Then

$$
\begin{aligned}
B * C &= B * A * B * C && (B * A * B = B) \\
&= B * B * C * A && (A\text{-movement}) \\
&= B * B * C * A * B * A && (A * B * A = A) \\
&= B * B * C * A * B * A * B * A && (B * A * B = B) \\
&= A * B * A * B * A * C * B * B && (A\text{-movement}) \\
&= A * B * A * C * B * B && (B * A * B = B) \\
&= A * C * B * B && (A * B * A = A) \\
&= C * B * A * B && (A\text{-movement}) \\
&= C * B && (B * A * B = B)
\end{aligned}
$$

CHAPTER 10

Combinatorics and Graph Theory

10.1 Graphs and Networks

125. A Campus Stroll

From the campus layout it is clear that the number of steps between two visits to the same building must be even. In Bob's sequence there are 30 steps between A and A, 16 between B and B, and so on to 14 steps between P and P, with two exceptions: 5 steps between J and J and 17 between K and K. Therefore J and K are the transposed buildings.

Using the corrected string, there are 8 possible labellings of the buildings, depending on which of the 4 corners is the degree-2 building A and which of A's 2 neighbors is O.

Note. There are several somewhat more complicated ways to attack this problem. The neat parity argument given here is due to Macalester student Mark Abbott-Cabezal.

126. Visit these Places

Observe that each vertex can be colored black or white so that adjacent vertices get different colors (see diagram; such a 2-colorable graph is called *bipartite*). Moreover, 7 vertices get one color and 9 get the other. Because a path as requested (called a *Hamiltonian path*) must alternate colors, it follows that no such path can exist.

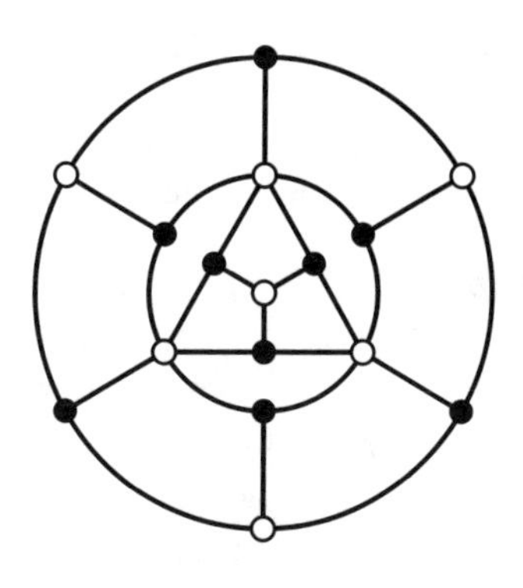

FIGURE 77

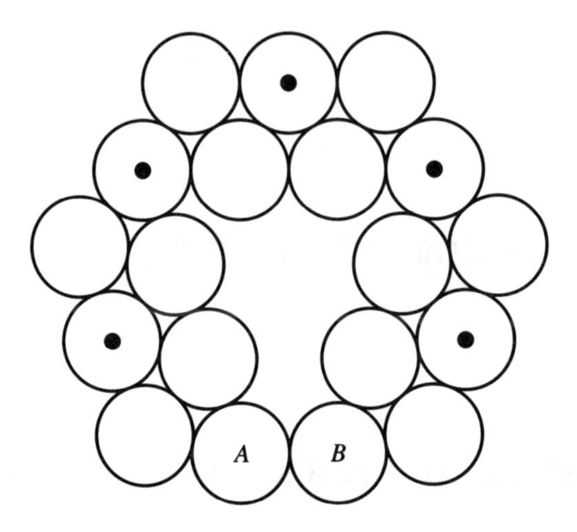

FIGURE 78

127. The Color of Money

The arrangement in Figure 78 requires four colors.

Suppose this arrangement could be colored with only three colors. Then starting with the penny labeled A and proceeding clockwise, it is not hard to see that all the pennies with dots in them must be the same color as A. But then penny B must also be the same color as A, violating the requirement that touching pennies be different colors.

128. Playing with Matches

Figure 79 gives a solution with 8 vertices; it is the smallest example.

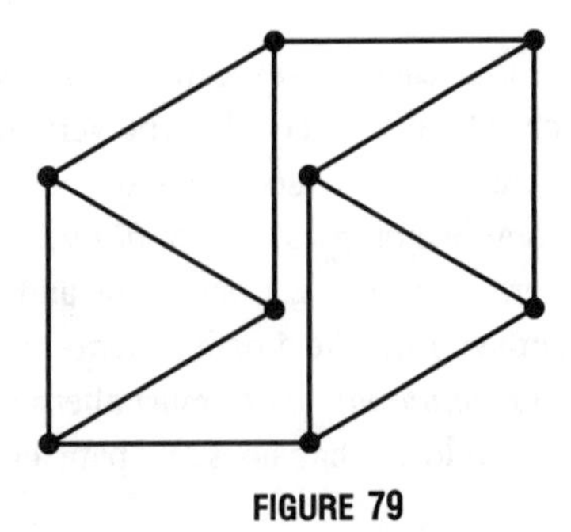

FIGURE 79

Problem 128.1. Let's relax the conditions and allow the sticks to cross. Then show that, for every d, there is a graph of this sort in which each vertex is adjacent to exactly d other vertices. In the case of $d = 4$, this can be done with 9 vertices.

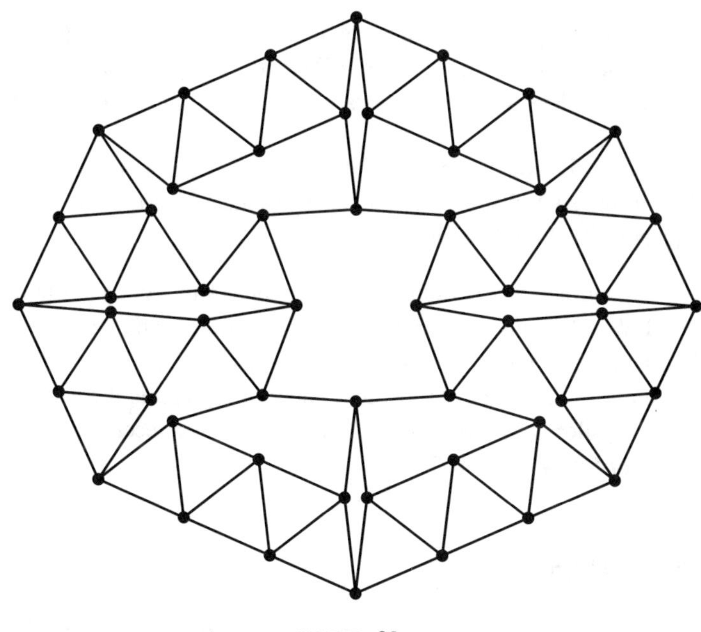

FIGURE 80

Notes. A graph in which each vertex has d neighbors is called regular of degree d. Thus Problem 128 asks for a matchstick graph that is regular of degree 3. In fact, it is possible (P. C. Fishburn and J. A. Reeds [FR]) to find such a graph having 20 vertices and with the property that the vertices form a convex 20-gon. It is known that there do not exist matchstick graphs that are regular of degree 5 or greater (see [Har1]). Heiko Harborth discovered a matchstick graph that is regular of degree 4 (see Figure 80); it has 52 vertices, the current best as far as minimizing the number of vertices goes.

Problem 128.1 is a variation of the following question of P. Erdős and G. Purdy. What is the minimum number of points in the plane such that each point has distance 1 to exactly d of the points? If $d = 1$, 2, 3, 4, or 5, the answers are known and are 2, 3, 6, 9, and 18, respectively (H. Harborth [Har2]).

If one allows edges to be formed by laying matchsticks end-to-end, an intriguing question arises: Can all planar graphs be drawn in the plane so that edges do not cross and each edge is a straight line of integer length? This problem, due to Harborth, is unsolved. It is a nontrivial exercise to find such a representation of K_4, the graph consisting of 4 vertices and all possible edges. (Hint: It can be done with the six edges having lengths 17, 17, 16, 10, 10, and 9; see [HR, §8.4].) Specializing to regular graphs, one can look for graphs that are regular of degree d and whose edges are noncrossing straight lines of integer length; several nice examples are given in [Har1]. If edges are allowed to cross (but they cannot lie in the same line), then it is known (see [HK, §2]) that every graph can be drawn in the plane such that all edges are straight lines of integer length. Here is an example.

Problem 128.2 (H. Harborth and A. Kemnitz [HK]). Find 4 points in the plane with no 3 in a line such that each has integer distance from the three other points and the largest such distance is 4.

Finally, we mention a very famous unsolved problem in the area of unit-distance graphs (see [KW]). What is the chromatic numbers of the graph whose vertex set is the set of *all* points in the plane and with two vertices being adjacent if they are unit distance from each other? The chromatic number of a graph is the least number of colors needed to color the vertices so that adjacent vertices receive different colors. The answer is one of 4, 5, 6, or 7.

129. A Universal Coloring

Consider a directed graph with one vertex for each color and all possible edges, including loops from a vertex to itself (see diagram). The coloring we seek would be provided by (and would provide) a circuit in this graph that starts and finishes at the same point and traverses each directed edge once; such a circuit is called an *Eulerian circuit.* It is well known, and not hard to prove, that such a path exists in a connected directed graph if and only if the in-degree of each vertex equals the out-degree. For the given example, with the obvious abbreviations, an Eulerian path is RGBYRBRYBBGYYGGRR.

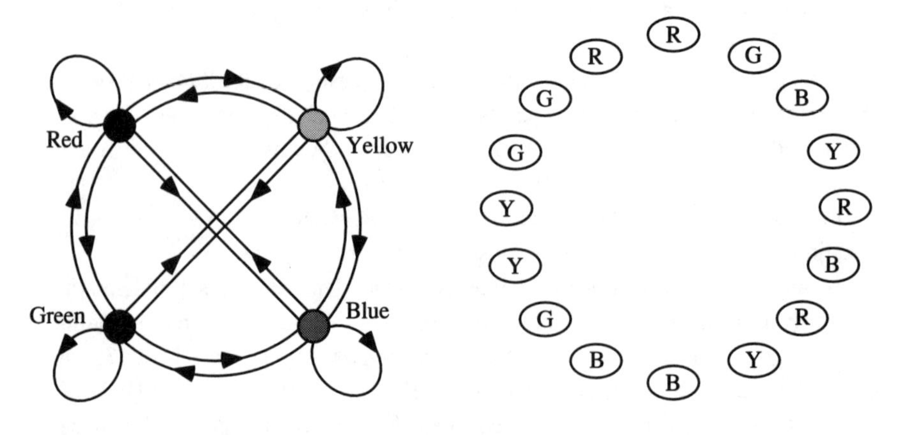

FIGURE 81

This argument shows that a "universal cycle for 2-sequences" exists for any collection of n colors; the cycle has length n^2. The same reasoning shows that a universal cycle for unordered pairs of four colors does not exist. In fact, a universal cycle for unordered pairs using n colors exists if and only if n is odd.

Notes. A universal cycle for k-sequences exists for any collection of n colors; a proof for $n = 2$ appears in [BM, §10.5], but it readily adapts to larger n. For unordered sequences (multisets) the question is more complicated. See [Jac] for a comprehensive discussion of results in ths area.

130. Hamiltonian Circuitry

No, it does not. To see why, first number the vertices in order from 1 to $2n$. If an edge connects vertex i to vertex j, we will say that its *direction number* is $i+j$. It is not hard to see that two edges are parallel if and only if their direction numbers are congruent modulo $2n$. Now suppose there were a Hamiltonian circuit in which no two edges were parallel. Then the direction numbers of the $2n$ edges in this circuit would have to be congruent to the numbers $1, 2, \ldots, 2n \pmod{2n}$, so their sum would have to be congruent to $1 + 2 + \cdots + 2n = 2n^2 + n \equiv n \pmod{2n}$. But since each vertex is an endpoint of two edges of the Hamiltonian circuit, the sum of the direction numbers is equal to twice the sum of the numbers of all the vertices, which is $2(1 + 2 + \cdots + 2n) = 4n^2 + 2n \equiv 0 \not\equiv n \pmod{2n}$. Thus we have reached a contradiction, so there can be no such Hamiltonian circuit.

Note. For a group-theoretic proof of this result, see [Sza].

131. Cover the Honeycomb

(a) Considering the analogy with chessboards, we can call the array a board and, for the problem at hand, the markers can be viewed as rooks. Thus part (a) seeks the maximum size of a set of nonattacking rooks (the independence number of the rook graph) on a triangular board.

The largest set of nonattacking rooks for the 10-rowed board is 7 (see Figure 82). The tough part is to prove this is the largest possibility, something that is generally very difficult for independence numbers of graphs. However, the particular case posed can be handled by the following argument, due to Jim Guilford (Digital Equipment Corporation, Hudson, Mass.). Suppose M, a maximal set of nonattacking rooks, has size k. Define the *value* of a cell to be the number of rooks in the three rows through the cell, not counting the cell itself. This number is at most 3 (otherwise M would have two rooks in one row). Moreover, the value of

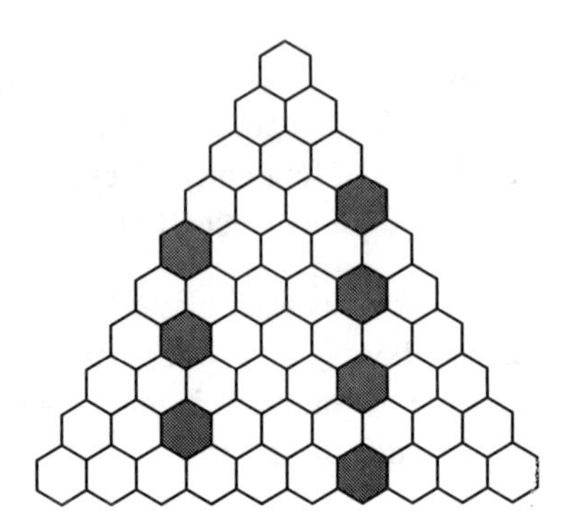

FIGURE 82

an occupied cell is 0 and the value of an empty cell is at most 3. It follows that the total value of all cells is no greater than $3(55 - k)$. Yet the actual total value must equal $18k$ because each rook sees 18 cells in the three directions. Therefore $18k \leq 3(55 - k)$, which tells us that $k \leq 7.86$, and so 7 is the largest possible size of an independent set.

(b) The answer is 5. To prove this we need to generalize to a board with n rows and handle the cases $n = 2$, 4, 6, 8, and 10, proving that the answers are 1, 2, 3, 4, and 5, respectively. Figure 83 shows the winning configurations for $n = 4$, 6, 8, and 10.

The case $n = 2$ is trivial; and it is clear that the $n = 4$ case cannot have a solution of size 1. For $n = 6$, we must show that 2 rooks cannot work. We may assume that neither of them is at one of the three corners, for otherwise we could use induction, appealing to the 4-row case after stripping off a vee-shaped formation. But then each row of length 6 must contain a rook, for otherwise at most 4 of the cells in such a row would be covered. (We call a cell *covered* if it is in the same row as

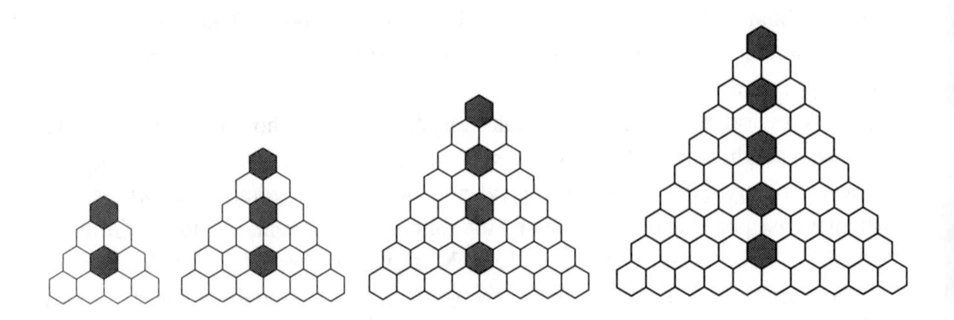

FIGURE 83

a cell containing a rook.) This is impossible, as there are three long rows but only two rooks.

On to $n = 8$. Suppose there is a solution of size 3. We again assume that no rook is in a corner. Then each row of size 8 and each row of size 7 must have a rook, because otherwise the entire row could not be covered ($3 \cdot 2 < 7$). There are 3 rooks and 6 such long rows, so each rook must do double duty. This can only be done by placing the three as in Figure 84, and then there are interior cells that are not taken care of.

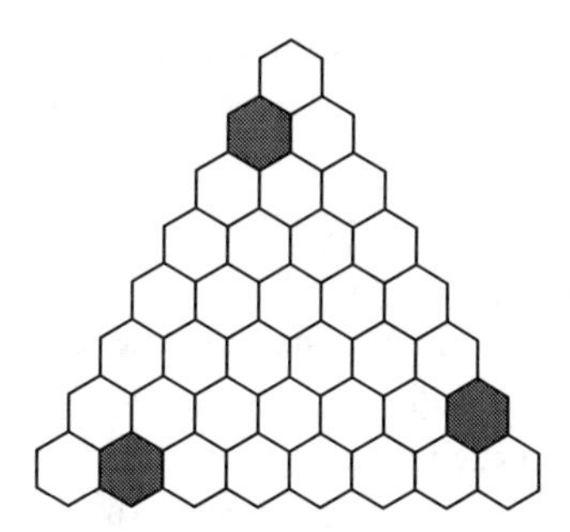

FIGURE 84

This brings us to $n = 10$. Suppose the array was covered with 4 markers. Assume none of these is in one of the 3 corners, for if one were, then an 8-row array would be covered by 3 markers, contrary to the proof just given.

Now, each of the 10-rows (rows of size 10) must contain a marker, for otherwise at most 8 of the positions in such a row would be covered. Similarly each of the 9-rows must contain a marker. There are therefore 6 large rows that need to be covered by 4 rooks, so that 2 of them must do double duty. Such double duty can be accomplished in two types of places: The positions 2 and 9 in the 10th row, and similarly in the tilted 10-rows, and positions 2 and 8 in the three 9-rows. If one of the latter is occupied, then that eliminates the other 9-row positions from double duty, so we can in any case assume that one of the 10-row positions that does double duty is occupied. Now there are two cases, according as the second double-duty point is in a 10-row or a 9-row. The latter is easiest, for if the already occupied 10-row position is in the bottom row and is position 9, then the 9-row position must be position 2 in the row above it. Then the row above that has 5 uncovered positions, so it must contain a marker, and that would be in position 8. But there are now 3 uncovered positions in rows 5, 6, 7 (locations 3, 3, 3, resp.) and this means the final rook cannot be placed in position 1, as it must to cover the uncovered 10-row. The other case is similar.

Notes. The question of what happens in the general n-row case is interesting. For part (a) Heiko Harborth (Braunschweig, Germany) has extended Jim Guilford's idea to derive a general formula for the independence number of the rook graph on the n-row board: it is $\lfloor (2n+1)/3 \rfloor$.

For part (b), it seems as if choosing rooks on an altitude of the triangle is best when n is even. But for odd n, Jim Guilford has shown that the altitude method is not optimal (except for $n = 1$, 3, or 5). The case $n = 7$ is central, for that admits a solution of size 3 (see Figure 85). By adding rooks at the apex, one can use the $n = 7$ solution to get a size-4 solution for $n = 9$, a size-5 solution for $n = 11$, and so on. Thus the pattern seems to be $1, 1, 2, 2, 3, 3, 3, 4, 4, 5, 5, 6, 6, 7, 7, 8, \ldots$. We have proofs in all cases where $n \leq 12$. For example, suppose $n = 9$ and a solution of size 3 existed. We may assume the corners are unoccupied (otherwise reduce to the $n = 7$ case). But then each 9-row, 8-row, and 7-row must have a marker. There are 9 such rows, but only 3 markers, contradiction, since triple duty is impossible.

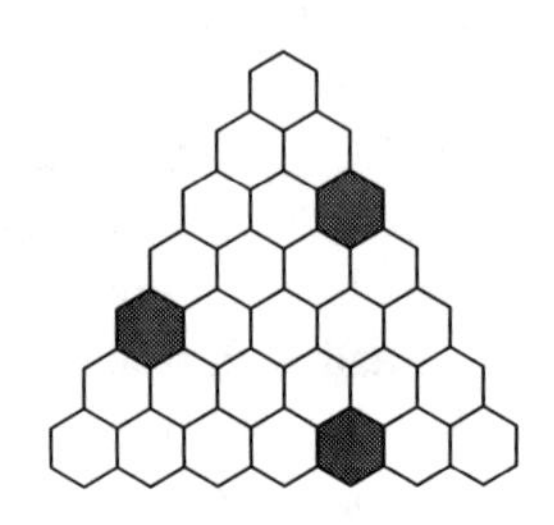

FIGURE 85

We now give Harborth's proof that $\lfloor (2n + 1)/3 \rfloor$ is the independence number. To see this is realized, one looks at examples similar to the one shown for $n = 10$ (further details of this are left to the reader).

To see that the formula is an upper bound, suppose we have r independent rooks on the board with n rows and $\binom{n+1}{2}$ cells. We can assume the set is maximal, and we must show that $r \leq (2n + 1)/3$.

Let the value of a cell be defined as before, and note that since each rook has $2n - 2$ empty cells in its lines, the total value of all cells is $(2n - 2)r$. We can decompose the value of a cell into the contributions from the three row directions. Define the horizontal value of a cell to be the number of rooks in the horizontal row through the cell, not counting the cell itself, and define the values for the other two directions similarly. Then the value of a cell is the total of its three directional values.

The horizontal value of a cell is 1 if it is in one of the r rows containing a rook but it is not itself occupied, and it is 0 otherwise. The number of cells with horizontal value 1 will be largest if the rows containing rooks are the r longest rows, so the total horizontal value of all cells is at most $(n - 1) + (n - 2) + \cdots + (n - r) = (2n - r - 1)r/2$. The same bound applies to the other two directions, so the total value of all cells is at most $3(2n - r - 1)r/2$. Thus $(2n - 2)r \leq 3(2n - r - 1)r/2$, which in turns yields $r \leq (2n + 1)/3$ as desired.

Many other problems of this type could be investigated. See [Fri] for a survey of analogous problems on the standard chessboard.

Problem 131.1 (H. Harborth). Determine a formula for the independence number of the king graph for a triangular board; in other words, find the maximum size of a set of nonattacking kings on a triangular board as in Problem 131. Note that kings attack only their immediate neighbors.

For Investigation. Determine the general solution to part (b) for $n > 12$.

10.2 Counting

132. Can You Get There From Here?

There are 21 routes from A to C. To see why, let x be the number of direct routes from A to B, that is, routes not passing through C. Let y be the number of direct routes from B to C, and let z be the number of direct routes from A to C. Then the given information tells us that $x + yz = 33$ and $y + xz = 23$. Adding and subtracting these equations gives us $(y - x)(z - 1) = 10$ and $(y + x)(z + 1) = 56$. Thus z is a positive integer such that $z - 1$ divides 10 and $z + 1$ divides 56, so the only possible values for z are 3 and 6. But solving for x and y leads to noninteger values of x and y when $z = 3$. Thus $z = 6$, from which we obtain $x = 3$ and $y = 5$, and the total number of routes from A to C is $z + xy = 21$.

133. Decimal Diversity

We need not consider numbers with 10 or more digits. Now, there are 9 ways of selecting the leading digit, and the rest is a permutation of some subset of the remaining digits. The number we seek is therefore

$$9\sum_{k=0}^{9}\binom{9}{k}k! = 9\sum_{k=0}^{9}\frac{9!}{(9-k)!} = 9\cdot 9!\left(\frac{1}{0!}+\frac{1}{1!}+\frac{1}{2!}+\cdots+\frac{1}{9!}\right).$$

Although it is easy to do the arithmetic and come up with the answer, 8877690, a more general approach will bring out an unexpected connection with the number e. Let s denote the sum of reciprocals just given; s is a partial sum of the standard series for e. Recall, by either Taylor's formula or a geometric series approximation, that $e - s < 1/9!$. Therefore $9!e - 1 < 9!s < 9!e$, so, because $9!s$ is an integer, $9!s = \lfloor 9!e \rfloor$, and so the answer, $9 \cdot 9!s$, is just $9 \cdot \lfloor 9!e \rfloor$, or 8877690. This means that the chance of a random integer between 1 and 10^{10} having distinct digits is about 1 in 1100.

134. Digit Counting

Make a list of positive integers by writing down, for each j in the sequence $1, 2, 3, \ldots, 10^n$, all the numbers you get by inserting a zero after any digit in j. For example, for $j = 274$, we would add the numbers 2074, 2704, and 2740 to the list. Clearly the number of entries added to the list for any value of j is the number of digits in j, so the length of the final list is the total number of digits in the sequence $1, 2, 3, \ldots, 10^n$. It is also clear that each number in the list is less than or equal to 10^{n+1}.

To establish the desired result we show that for any positive integer $k \leq 10^{n+1}$, the number of times that k appears in the list is equal to the number of zeros in k, and therefore the length of the list is also the total number of zeros in the sequence $1, 2, 3, \ldots, 10^{n+1}$. The reason for this is simply that for each zero in k, k is listed once with that zero being the added zero. For example, if $k = 200704$, then k appears three times in the list, twice from adding zeros to $j = 20704$ (adding zeros after both the first and second digits) and once from adding a zero to $j = 20074$.

One interesting feature of this solution is that we have shown that two numbers have the same value without actually finding that value! This suggests the following problem.

Problem 134.1. How many digits are there in the sequence $1, 2, 3, \ldots, 10^n$?

135. How Long Would It Take You?

We will solve the more general problem involving a lock with n buttons, for any positive integer n. Let a_n be the number of valid combinations for such a lock. It is most convenient in our calculations to include as a valid combination the "empty combination" (i.e., the combination in which no buttons are pushed). We can eliminate this combination again at the end of our calculations.

A valid combination for a lock with n buttons is either the empty combination, or else it consists of a collection of k buttons that are pushed first, for some k between 1 and n, followed by a valid combination (possibly empty) involving the remaining $n-k$ buttons. Since for each k there are $\binom{n}{k}$ ways to choose k buttons to be pushed first and a_{n-k} valid combinations involving the remaining $n-k$ buttons, we have

$$a_n = 1 + \binom{n}{1}a_{n-1} + \binom{n}{2}a_{n-2} + \cdots + \binom{n}{n}a_0.$$

This formula can be used to compute the a_ns recursively. To get the recursion started, note that for a lock with no buttons the empty combination is the only valid combination, so $a_0 = 1$.

We can implement this recursion in *Mathematica* using the following definitions (the "`a[n] = ...`" is used to store each `a` value as it is computed, thus avoiding an exponential blowup in the computation time):

```
a[0] = 1;
a[n_] := (a[n] = 1 + Sum[Binomial[n, k] a[n - k], {k, 1, n}])
```

Asking *Mathematica* for `a[5]` produces the answer 1082. Thus, discarding the empty combination, there are 1081 nonempty valid combinations for the five-button lock.

Notes. Of course, it would be nice to find a general formula for a_n by solving our recurrence relation. Since the binomial coefficients in the recurrence relation all involve a factor of $n!$, the problem can be simplified somewhat by defining $b_n = a_n/n!$. The recurrence for a_n then leads to the following recurrence for b_n.

$$\begin{aligned} b_0 &= 1. \\ b_n &= \frac{1}{n!} + b_{n-1} + \frac{b_{n-2}}{2} + \frac{b_{n-3}}{3!} + \cdots + \frac{b_0}{n!} \\ &= \frac{1}{n!} + \sum_{k=1}^{n} \frac{b_{n-k}}{k!}, \qquad \text{for } n \geq 1. \end{aligned}$$

Some remarkable formulas can now be derived by studying the function

$$f(x) = \sum_{n=0}^{\infty} b_n x^n.$$

Problem 135.1. Show that the series defining $f(x)$ converges absolutely for $|x| < \ln 2$ by showing that for all n,

$$\frac{1}{(\ln 2)^n} \leq b_n \leq \frac{2}{(\ln 2)^n}.$$

Substituting the recursion for b_n into $f(x)$, we find that

$$\begin{aligned} f(x) &= b_0 + \sum_{n=1}^{\infty} b_n x^n = 1 + \sum_{n=1}^{\infty} \left[\frac{x^n}{n!} + \sum_{k=1}^{n} \frac{b_{n-k} x^n}{k!} \right] \\ &= 1 + \sum_{n=1}^{\infty} \frac{x^n}{n!} + \sum_{n=1}^{\infty} \sum_{k=1}^{n} \frac{x^k}{k!} (b_{n-k} x^{n-k}) = e^x + \sum_{k=1}^{\infty} \frac{x^k}{k!} \sum_{n=k}^{\infty} b_{n-k} x^{n-k} \\ &= e^x + \sum_{k=1}^{\infty} \frac{x^k}{k!} \sum_{n=0}^{\infty} b_n x^n = e^x + (e^x - 1) f(x). \end{aligned}$$

Thus $(2 - e^x) f(x) = e^x$, so

$$f(x) = \frac{e^x}{2 - e^x}.$$

Since b_n is the coefficient of x^n in the Maclaurin series for $f(x)$, it follows that

$$b_n = f^{(n)}(0)/n!.$$

But recall that $a_n = n!b_n$, so

$$a_n = f^{(n)}(0) = \frac{d^n}{dx^n}\left[\frac{e^x}{2-e^x}\right]_{x=0}.$$

One way to find a formula for these derivatives is to rewrite $f(x)$ as a geometric series:

$$f(x) = \frac{e^x}{2-e^x} = \frac{e^x/2}{1-e^x/2} = \sum_{k=1}^{\infty}\left(\frac{e^x}{2}\right)^k.$$

Differentiating this series term-by-term we find that

$$f^{(n)}(x) = \sum_{k=1}^{\infty} k^n \left(\frac{e^x}{2}\right)^k,$$

so, finally, we have a succinct formula for the number of combinations:

$$a_n = f^{(n)}(0) = \sum_{k=1}^{\infty} \frac{k^n}{2^k}.$$

Thus $a_n = \mathrm{Li}_{-n}(\frac{1}{2})$, where Li is the polylogarithm function, which is defined as follows (see [Lew, pp. 189, 236]):

$$\mathrm{Li}_n(z) = \sum_{k=1}^{\infty} \frac{z^k}{k^n}.$$

For more on this problem, see [CV]. Many of the formulas discussed here can be derived by *Mathematica,* as is shown in [Abb].

136. How Many Triangles?

First observe that the number of nonadjacent pairs from a linear array of k objects is $\binom{k}{2} - (k-1)$, or simply $\binom{k-1}{2}$. It follows that the number of triangles containing a particular vertex is $\binom{n-4}{2}$, since this reduces to choosing a nonadjacent pair from an array of $n-3$ objects (the neighbors of the specified vertex are excluded). Summing over the n vertices and adjusting for the fact that each triangle is counted thrice, we get $\frac{n}{3}\binom{n-4}{2}$, or $n(n-4)(n-5)/6$. This formula is valid only for $n \geq 4$; there are of course no triangles of the type sought if $n \leq 3$. The first several values of the formula, starting from $n = 4$, are: 0, 0, 2, 7, 16, 30, 50, 77, 112, 156, 210.

137. Breaking Up Space

One can make a careful sketch and see that the answer to (a) is 14. But we can use induction to solve the general problem. Let $f(n)$ be the number of regions generated by n planes; $f(1) = 2$. Now, suppose we have a configuration of n planes through a common point so that no three of them meet in a line—Figure 86 shows 3 such planes—and a new plane through the common point is added so that it contains none of the lines that are intersections of the old planes. The new plane intersects the old planes in n distinct lines through the common point, thus dividing the new plane into $2n$ regions. Each of these regions divides an existing section of space into two. Therefore $f(n+1) = f(n) + 2n$. We can examine a few values:

$$f(1) = 2$$

$$f(2) = 2 + 2 \cdot 1$$

$$f(3) = 2 + 2 \cdot 1 + 2 \cdot 2$$

$$f(4) = 2 + 2 \cdot 1 + 2 \cdot 2 + 2 \cdot 3$$

$$f(5) = 2 + 2 \cdot 1 + 2 \cdot 2 + 2 \cdot 3 + 2 \cdot 4$$

This leads us to the conclusion that $f(n)$ is just the sum of 2 with the sum of the first $n - 1$ even integers. Therefore

$$f(n) = 2 + 2 + 4 + \cdots + 2(n-1) = 2 + 2(n-1)n/2 = n^2 - n + 2.$$

FIGURE 86

The sequence of f-values is therefore $2, 4, 8, 14, 22, 32, 44, 58, \ldots$. Note that this formula does not work in the case of 0 planes, for then the correct answer is 1, while the formula gives 2.

Alternate Solution. Consider the general problem with n planes and imagine a sphere centered at the common point. The planes divide the spherical surface into regions called faces, which are in one-to-one correspondence with the regions of space we are counting. We can apply Euler's formula for connected configurations (graphs) on a sphere; it says that $V - E + F = 2$, where V, E, and F are the number of vertices, edges, and faces, respectively. The configuration here is the graph on the sphere determined by the intersections of the sphere with the lines common to pairs of planes, with the edges being the arcs in which the planes intersect the sphere. Each pair of planes determines a line that hits the sphere in two points, so $V = 2\binom{n}{2}$. Each vertex has edges (arcs) to four other vertices on the sphere (two on each of the two planes through it); thus the number of edges is $4V/2$, since $4V$ counts each edge twice. The equation is then easily solved to get $F = n^2 - n + 2$.

Problem 137.1. Into how many regions do n circles divide the plane, if each two circles intersect in two points, and no three of the circles pass through the same point?

The induction technique works, but Euler's formula, valid for connected graphs in the plane, is quicker. We have $n(n-1)$ vertices, since each circle has $2(n-1)$ points but this counts each point twice, and $2n(n-1)$ edges, since each circle is divided into $2(n-1)$ arcs. This data, combined with $V - E + F = 2$ leads to $n^2 - n + 2$ regions, the same answer as for the plane problem in 3-space.

In fact, there is a geometrical connection between the plane and circle problems that shows why the number of regions is the same. Start with the sphere of the first solution above and assume the north pole is not on one of the great circles made by the planes. Then project the sphere from the north pole onto the horizontal plane tangent to the south pole. This projection sends circles on the sphere onto circles in

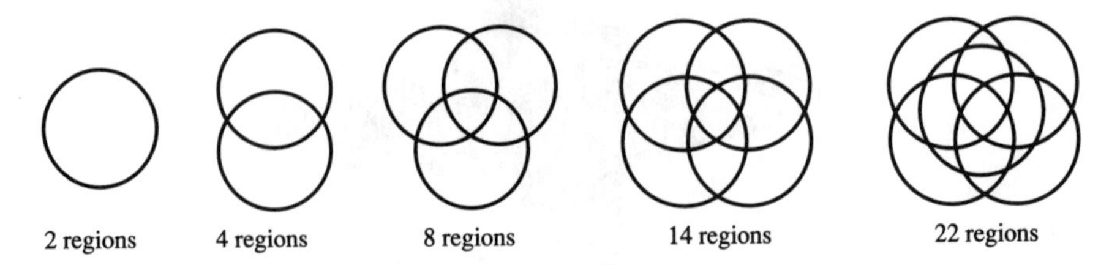

FIGURE 87

the plane (proof omitted) and the regions into which the circles divide the plane are in one-to-one correspondence with the regions of 3-space obtained from the planes.

138. A Piercing Diagonal

The number of squares containing a segment of the diagonal is $m+n-\gcd(m,n)$, where $\gcd(m,n)$ is the greatest common divisor of m and n.

Suppose first that $\gcd(m,n)=1$. Then the diagonal does not pass through any corners of grid squares, so when it passes from one square to another it always crosses either a vertical or a horizontal line. Since it must cross all of the $m-1$ vertical and $n-1$ horizontal lines, there will be a total of $m+n-2$ crossings, and therefore the line will pass through $m+n-1$ squares.

If $\gcd(m,n)=d\neq 1$ then $m=dm'$ and $n=dn'$ for some relatively prime integers m' and n'. In this case the diagonal can be broken into d pieces, each beginning and ending at a corner of a square and going m' squares horizontally and n' squares vertically. We already know that each of these pieces will pass through $m'+n'-1$ squares, so the whole diagonal will pass through

$$d(m'+n'-1)=dm'+dn'-d=m+n-\gcd(m,n)$$

squares.

Problem 138.1. A paintbrush whose width is the same as the width of a grid square is used to paint a line from one corner of the grid to the opposite corner. How many squares will have paint in their interiors?

139. Can You Find the Key?

Consider the coloring:

RED GREEN RED RED GREEN GREEN GREEN GREEN GREEN GREEN … .

This works for a set of n keys provided $n\geq 6$; each key in the string of GREENs is identifiable by its distance from the isolated RED. Thus $f(139)=2$. It is not hard to see that $f(4)=f(5)=3$. So the entire sequence of f-values looks like: $1,2,3,3,3,2,2,2,2,2,\ldots$.

Note. This question was posed by Frank Rubin [Rub]. The problem was used in the 1984 Canadian Mathematical Olympiad [Doo]. Michael Albertson and Karen Collins [AC] have investigated variations of this question where the key configuration forms a graph other than a cycle, which is the graph underlying a round key ring. Their main result is that two colors suffice for graphs having an Abelian automorphism group.

140. Count the Tilings

Let a_n be the number of tilings of a $2 \times n$ grid using 1×1 and 1×2 tiles so that the two rightmost squares are occupied by singletons, let b_n be the number of tilings with one singleton and one doubleton in the rightmost squares, and let c_n be the number of other tilings (two horizontal doubletons or one vertical doubleton). The problem asks for $a_7 + b_7 + c_7$. When one appends another column to the right side, forming a $2 \times (n+1)$ grid, either one can add 2 singletons or a vertical doubleton, or one can change any singletons in the nth column to doubletons. This yields three recurrence relations:

$$a_{n+1} = a_n + b_n + c_n$$

$$b_{n+1} = 2a_n + b_n$$

$$c_{n+1} = 2a_n + b_n + c_n$$

Since $a_1 = 1$, $b_1 = 0$, and $c_1 = 1$, we can easily generate a table of the values that yields the answer, 2356.

n	a_n	b_n	c_n	sum
1	1	0	1	2
2	2	2	3	7
3	7	6	9	22
4	22	20	29	71
5	71	64	93	228
6	228	206	299	733
7	733	662	961	2356

Notes. The solution presented above is due to John Guilford (Hewlett-Packard, Everett, Wash.). Let d_n denote the number of tilings of a $2 \times n$ grid using 1×1 and 1×2 tiles (note that d_n, being the sum of a_n, b_n, and c_n, is just a_{n+1}). Then d_n satisfies the recurrence $d_n = 3d_{n-1} + d_{n-2} - d_{n-3}$. Problem 140 is due to Brigham, Chinn, Holt, and Wilson. Their paper [BCHW] contains six different approaches to the problem of finding the recurrence for d_n. They also show that d_n can be expressed in closed form, though it is quite complicated: $d_n =$

$c_1 r_1^n + c_2 r_2^n + c_3 r_3^n$ where $r_1 = (4/\sqrt{3})\cos\alpha + 1$, $r_2 = -(2/\sqrt{3})\cos\alpha - 2\sin\alpha + 1$, $r_3 = -(2/\sqrt{3})\cos\alpha + 2\sin\alpha + 1$, and $\alpha = \frac{1}{3}\arctan(\sqrt{111}/9)$; the constants c_1, c_2, c_3 are determined from the three initial conditions: $d_0 = 1$, $d_1 = 2$, $d_2 = 7$. (Note that $d_0 = 1$ because there is only one way to tile the empty grid: do nothing.) The symbolic forms of the three constants c_i are horrendous, though a symbolic computation package can find them very quickly. Numerically, they are: $c_1 = 0.664591\ldots$, $c_2 = 0.255972\ldots$, and $c_3 = 0.0794367\ldots$. For other problems of this type see [BCCG] and [CCFM].

Problem 140.1 [BCCG]. How many tilings are there of a $1 \times n$ grid using singletons and dominoes?

Problem 140.2. Let d_n denote the number of tilings of a $2 \times n$ grid using 1×1 and 1×2 tiles. Prove that d_n satisfies the recurrence $d_n = 3d_{n-1} + d_{n-2} - d_{n-3}$.

Problem 140.3 [CCFM]. Look at all the tilings of a $1 \times n$ grid using tiles that are any of $1 \times 1, 1 \times 2, \ldots, 1 \times n$, and let W_n be the number of singletons that occur on all these tilings. Then $W_2 = 2$ and $W_3 = 5$. Find a recurrence that expresses W_n in terms of prior values of W.

Problem 140.4 [GKP, §7.1, 7.3]. Find a method for computing the number of tilings of a $3 \times n$ grid using dominoes only.

10.3 Miscellaneous

141. How to Win at the Lottery

Let a, b, c, d, e, f be the lucky numbers chosen by the lottery commission and suppose u is the remainder when $a + b + c + d + e + f$ is divided by 47. If $u = 0$, we own the ticket and we are big winners! Otherwise, we must show that we can change one of a, b, c, d, e, f to a number between 1 and 48 such that (a) it is not one of a, b, c, d, e, f, and (b) the change makes the sum divisible by 47. For then the changed ticket will be one we own and will agree with the lucky ticket in 5 places (the untouched ones).

In order to accomplish this change, we must either decrease the sum by u (a "move down") or increase it by $47 - u$ (a "move up"); note that u is between 1 and 46. Observe that for any integer r in $[1, 48]$ at least one of $r - u$ or $r + (47 - u)$ is also in the interval. Thus we start by replacing a with either $a - u$ or $a + (47 - u)$, whichever is in $[1, 48]$. This works unless the new number is one of the others, say b. In that case we don't change a, but replace b with either $b - u$ or $b + (47 - u)$.

If this approach does not work in any of the six cases, then a sequence of six steps starting from, say, a would visit b, c, d, e, and f and return to a. (Actually, there is the possibility of a shorter loop, perhaps of the form ($a \to d \to c \to f \to d$), but that is taken care of by the argument that follows, starting from d.) If i is the number of up steps in these 6 steps, and j the number down, then $a + i(47 - u) - ju = a$, or $(i + j)u = 47i$. Because 47 does not divide u, this implies that 47 divides $i + j$, which is impossible since $1 \leq i + j \leq 6$.

Problem 141.1. How many of the $\binom{48}{6}$ tickets have their sum divisible by 47?

Problem 141.2. Find a winning scheme that is better than the one in Problem 141 in that it requires the purchase of fewer tickets.

Notes. Problem 141 was raised by David Kelly (Hampshire College). It has been considered in the literature [Liu3], but very little is known about the general question. The general question of the distribution of subset sums modulo m (as in Problem 141.1) is discussed in detail in [WW], where it is proved that the t-subsets of $\{1, \ldots, n\}$ are equidistributed with respect to modulo-m-sums if and only if $t \bmod d > n \bmod d$ for each d that divides m (except $d = 1$; "$a \bmod b$" here denotes the least nonnegative residue of a modulo b).

142. A Problem with the Elevators

If there is any floor on which at most two elevators stop, then the building can have at most $5 + 5 + 1 = 11$ floors. If at least three elevators stop at each floor, then there are at most 14 floors, since

$$(7 \text{ elevators}) \times (6 \text{ stops per elevator})/3 \text{ stops per floor} = 14 \text{ floors}.$$

To see that an arrangment serving 14 floors exists, consider a geometrical arrangement of 7 "lines" and 7 points arranged so that each line has 3 points, each point is on 3 lines, each pair of points is connected by one line, and each pair of lines intersects in one point. A geometrical view of such an object—called a projective plane of order 2—is given in the diagram (the circle is considered to be a line). Now

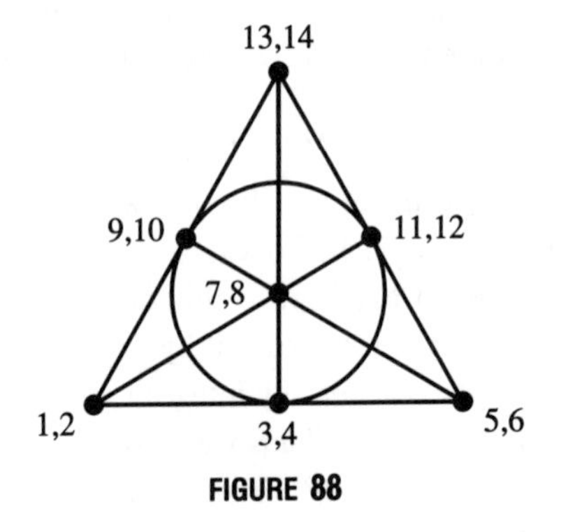

FIGURE 88

label the points with pairs of floors and let the lines be elevators. This yields the following 14-floor solution:

Elevator number 1:	1	2	3	4	5	6
Elevator number 2:	1	2	7	8	11	12
Elevator number 3:	1	2	9	10	13	14
Elevator number 4:	3	4	7	8	13	14
Elevator number 5:	5	6	7	8	9	10
Elevator number 6:	3	4	9	10	11	12
Elevator number 7:	5	6	11	12	13	14

Note. This problem is due to Andy Liu [Liu1].

143. Avoiding Arithmetic Progressions

The integers between 1 and 26 may be decomposed as follows to solve the problem:

$$\{1, 2, 5, 6, 12, 14, 15, 17, 21\}$$
$$\{3, 4, 7, 9, 16, 18, 19, 24, 26\}$$
$$\{8, 10, 11, 13, 20, 22, 23, 25\}.$$

If we code the splitting above by the string $s_1 = 00110012122020010112022121$ (the digits tell which of the three sets to use for the integers from 1 to 26), then other splittings are

$$s_2 = 00112002021122112020021100$$

$$s_3 = 00121002021122112020012100$$

$$s_4 = 01001012211221010012200202$$

$$s_5 = 21001012211221010012200202$$

The strings s_2 and s_3 are palindromes, but the other three can be reversed to yield different splittings, so that there are eight solutions in all (where the order of the three sets is ignored). See [BO] for more along these lines, along with a discussion of the algorithms that can be used to find arithmetic-progression-free strings.

Notes. A theorem of van der Waerden states that, for every p, there is a number $W(p)$ such that whenever $n \geq W(p)$ and $\{1, \ldots, n\}$ is split into p sets, one of the sets has a 3-term arithmetic progression. As shown in [BO], $W(3) = 27$. Since $W(1) = 3$ and $W(2) = 9$, one might guess that $W(4) = 81$. But in fact $W(4) = 76$, a result that required 1200 hours of computing time in 1979.

Problem 143.1. Find the two other ways of splitting $\{1, 2, \ldots, 8\}$ into two sets each having no 3-term arithmetic progression.

Problem 143.2. Show that such a splitting of $\{1, 2, \ldots, 9\}$ does not exist, thus proving that $W(3) = 9$.

144. Easy as One, Two, Three

Suppose $\binom{n}{k}$, $\binom{n}{k+1}$, $\binom{n}{k+2}$ are in the ratio $1:2:3$. Then

$$2 = \frac{k!(n-k)!}{(k+1)!(n-k-1)!} = \frac{n-k}{k+1}$$

and

$$\frac{3}{2} = \frac{(k+1)!(n-k-1)!}{(k+2)!(n-k-2)!} = \frac{n-k-1}{k+2},$$

so we obtain $2n - 5k = 8$ and $n - 3k = 2$. Solving gives $n = 14$ and $k = 4$; thus the unique triple of consecutive binomial coefficients in the ratio $1:2:3$ is $\binom{14}{4}$, $\binom{14}{5}$, and $\binom{14}{6}$, or 1001, 2002, and 3003.

Note. This problem is due to C. W. Trigg [Tri1]. Note also that because $\binom{14}{7} = 3432 \neq 4004$, there is no sequence of consecutive binomial coefficients that is in the ratio $1:2:3:4$.

145. Difference Triangles

Let n denote the number of rows. Here is a complete list of solutions for $n = 2$, 3, 4, and 5. Any solution yields another by a flip about the vertical center line, and so we list only half the solutions. Moreover, we need give only the top row, since the rest follows.

$n=2$	$\{1,3\}$	$\{2,3\}$		
$n=3$	$\{1,6,4\}$	$\{4,1,6\}$	$\{2,6,5\}$	$\{5,2,6\}$
$n=4$	$\{6,1,10,8\}$	$\{6,10,1,8\}$	$\{8,3,10,9\}$	$\{8,10,3,9\}$
$n=5$	$\{6,14,15,3,13\}$			

Some techniques for minimizing search time are discussed by C. W. Trigg [Tri3]; for example, if $n = 5$, 15 must occur in the top row, 13 and 14 must occur in the top two rows, 10, 11, and 12 must occur in the top three rows, and also $\{1, 2, 3, 4, 5\}$ must occur on different rows.

FIGURE 89

However, there is no difference triangle for $n = 6$. In the proof that follows, "small triangle" means an array of three positions in the shape of an equilateral triangle with a horizontal upper boundary. Now, if there were a difference triangle for $n = 6$, it would use the integers from 1 to 21, and so the parity of the sum of the labels (231) is odd. But the parity of the sum of the entries in a difference triangle with 6 rows must be even. To prove this observe that the parity of any small triangle (3 positions) is even (odd − odd = even, odd − even = odd, even − even = even). Now cover a 6-rowed difference triangle with small triangles by taking the three that contain the corners together with all the small triangles that do not intersect these three (see Figure 89). This collection of triangles covers each position either once or three times. This means that the overall parity of the triangle is even, as claimed. This argument extends by induction to all n of the form $2^k - 2$ ($k \geq 3$). For example, to go from 6 to 14, cover the three corner areas with 6-rowed triangles and consider all small triangles that do not intersect these. This shows that the overall parity is even.

Notes. Computer searches have shown (see [Tri3], [Gar1, pp. 119–120, 128–129]) that there are no difference triangles with 7 or 8 rows using $[1, 28]$ or $[1, 36]$, respectively, and that the 5-row solution is unique. It has been conjectured that there are none for any larger number of rows. This problem is discussed by Martin Gardner [Gar1, p. 128], who presents a parity proof due to Sicherman that proves nonexistence of difference triangles when the number of rows has the form $2^n - 2$. The proof given here, which is a little different than Sicherman's, was found by Curtis Greene (Haverford College).

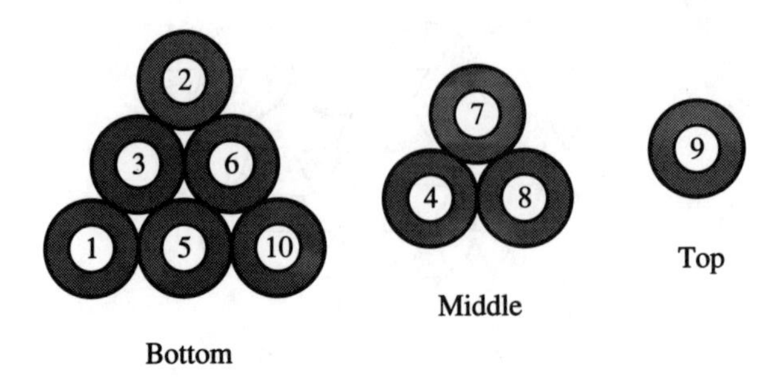

FIGURE 90

146. Ten Cannonballs

Note that the corner balls (*estranged*) have three neighbors, while the others (*gregarious*) have six. If either of 1 or 10 is gregarious then a difference of 6 or more must show up. If both 1 and 10 are estranged, then they have a common neighbor, which implies that a difference of 5 or more must show up. Thus 5 appears to be optimal. The array in the diagram shows that 5 is indeed the minimum.

147. Christmas Confusion

Here is a 26-move solution. Because cards must move into the space, it is sufficient to list the cards in the order in which they are moved.

> to You Dept. the Merry Dept. the Math from Merry Math from Merry Xmas Dept. Math from the You Dept. Math to Dept. You to Dept.

For Investigation. Is 26 the smallest number of moves for which a solution exists?

148. A Veritable Babel

The answer is 10. Since $\binom{10}{5} = 252$, each student could speak a different set of 5 languages chosen from a fixed list of 10 languages. Clearly the hypotheses of the problem would be satisfied in this situation, so 10 languages suffice. To see that 10 is minimal, we use the following lemma.

Lemma. *Suppose F is a family of subsets of $\{1, 2, \ldots, n\}$ with the property that no set in F is a subset of any other set in F. Then F has at most $\binom{n}{\lfloor n/2 \rfloor}$ elements.*

Proof. Let us say that a set X in F is compatible with a permutation $i_1, i_2, \ldots, i_n$ of the set $\{1, 2, \ldots, n\}$ if $X = \{i_1, i_2, \ldots, i_j\}$ for some $j \leq n$. Clearly if X has

j elements then X is compatible with $j!(n-j)!$ different permutations. It is not hard to see that $j!(n-j)! \geq k!(n-k)!$, where $k = \lfloor n/2 \rfloor$, so every set in F is compatible with at least $k!(n-k)!$ different permutations.

Furthermore, if two sets X and Y were compatible with the same permutation then one of them would be a subset of the other, a contradiction; so no two elements of F can be compatible with the same permutation. But there are only $n!$ permutations altogether, so the number of elements of F is at most $n!/(k!(n-k)!) = \binom{n}{\lfloor n/2 \rfloor}$, proving the lemma.

Now, to see that 10 is minimal for our problem, suppose that the languages spoken by the students are $L_1, L_2, \ldots, L_n$. For each student A let $X_A = \{i\colon A \text{ speaks } L_i\}$, and let F be the family of all the sets X_A. Then F has 250 elements, and it is easy to see that F satisfies the hypotheses of the lemma. Therefore by the lemma $\binom{n}{\lfloor n/2 \rfloor} \geq 250$. But $\binom{9}{4} = 126$, so $n \geq 10$.

Notes. This problem appeared on the 1992 USA Mathematics Talent Search and appeared in [Ber1, problem 4]. The lemma was first proven by Sperner [Spe]; the proof we have given is due to D. Lubell [Lub].

149. Find the Pattern

The answer is 2001. Observe that the 2×2 box at the lower left has each of 0 and 1 in each row and column. That means that the 2×2 box directly above it will be identical in structure, but will begin with 2 instead of 0; the same holds for the 2×2 box to the right. And the 2×2 up and right will then be identical to the lower left one, since none of 0 or 1 appears to the left or below. This means that the lower left 4×4 box will have each of 0, 1, 2, and 3 in each row and column, and so we can repeat the same argument. This gives a straightforward algorithm for determining any value. For a given pair (a, b) look at the smallest $2^n \times 2^n$ box that contains it and divide it into four quadrants. By minimality (a, b) will not be in the lower left of these four quadrants; move the point to the corresponding position in the lower left quadrant. If the point was at the upper right, this causes no change in its value. Otherwise, this subtracts 2^{n-1} from its value. This process can be repeated until we are at the $(0, 0)$ position.

We can use base-2 digits to keep track of the quadrants that show up. Write both coordinates in base-2 notation, filling with zeros so that the representations have the same length. If the leading digits are different, then a power of 2 must be accounted for. Otherwise, the leading digits can be removed, causing no change in the values of the mystery function. The final answer is therefore the sum of the powers of 2 that show up as we work our way down. This is simply the exclusive-or function of the two binary strings. We illustrate with $1707 = 11010101011_2$ and $378 = 00101111010_2$:

$$\begin{array}{r} 11010101011 \\ 00101111010 \\ \hline 11111010001, \end{array}$$

and 11111010001 is the base-2 representation of 2001.

The preceding arguments can be made more rigorous by appealing to the fact that the nonnegative integers form a group under $\oplus$, the base-two exclusive-or function. For a number n, let n_i be its ith binary digit, that is, the coefficient of 2^i in its binary expansion. We are grateful to George Bergman (University of California, Berkeley) for the proof that follows.

Proposition. *If $k < m \oplus n$ then either $k = m \oplus n'$ for some $n' < n$ or $k = m' \oplus n$ for some $m' < m$.*

Proof. Let d be the largest index i such that k_i and $(m \oplus n)_i$ disagree. Then $(m \oplus n)_d = 1$ and $k_d = 0$, since $k < m \oplus n$. Then, say, $m_d = 1$ and $n_d = 0$. Let m' be the unique number such that $m' \oplus n = k$. Then $(m' \oplus n)_i$ agrees with $(m \oplus n)_i$ for $i > d$, so m'_i agrees with m_i for $i > d$, and $m'_d = 0$ since $k_d = n_d = 0$. Hence m' is less than m.

The proposition, together with induction to know that all earlier entries follow the $\oplus$-rule, shows that the number in the (m, n)th position is at least as large as $m \oplus n$. Equality then follows from the facts that $m \oplus n \neq m \oplus n'$ for $n' < n$ and $m \oplus n \neq m' \oplus n$ for $m' < m$. These last are true because $\oplus$ is a group operation.

Note. The pattern can also be described in terms of the game Nim: the entry in the position (m, n) is the Nim-sum of m and n. For more on the connection with games, as well as an interpretation in terms of fields see [Con, chap. 6].

CHAPTER 11

Three-Dimensional Geometry

150. Only Isosceles Triangles

Start with a regular pentagon inscribed in the unit circle in the x-y plane, together with the origin; this places six of the points. Let the remaining two be $(0, 0, \pm 1)$. It is easy to see that the resulting figure, which looks like two Chrysler logos placed back-to-back, solves the problem. There are seven types of isosceles triangles in this configuration (as well as the degenerate triangle consisting of three collinear points).

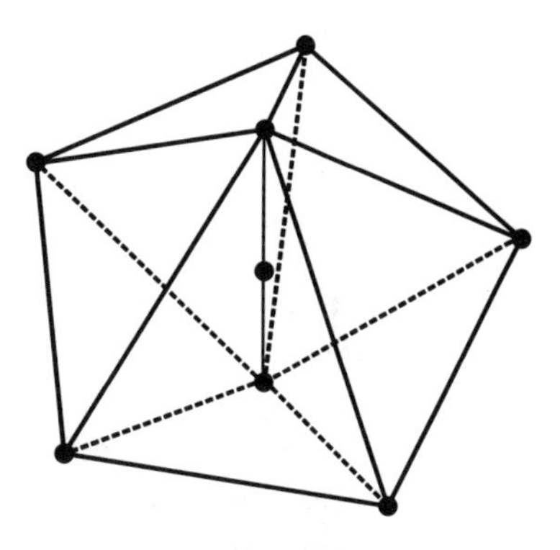

FIGURE 91

Notes. This problem was posed by Paul Erdős ([Erd2]; see also [CFG, problem F6] and [Ran]) and solved by L. M. Kelly. In addition to finding the 8-point "isosceles set" that solves this problem, Kelly proved that the only 6-point isosceles set in the plane is the regular pentagon together with its center (see [Hon, p. 123–132]). Croft [Cro] has proved that there are no 9-point isosceles sets in 3-space; it is not known whether the 8-point set of the solution is unique.

151. Finding the Middle Ground

The midset of S and T is the square connecting the centers of the faces not containing S and T; see Figure 92. One approach to a proof is to show that the midset of P and T, where P is a point on S, is a slice of the square.

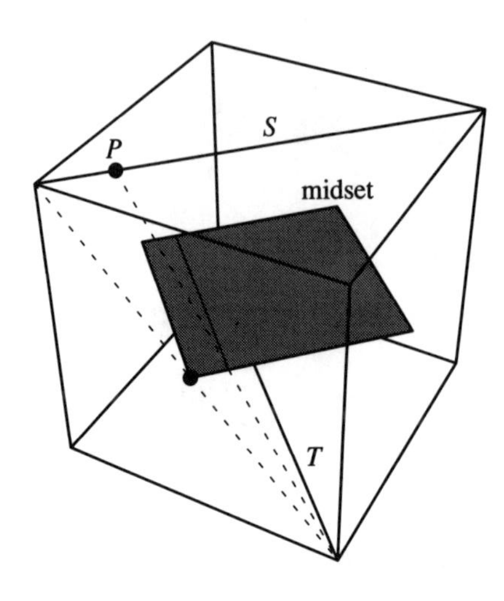

FIGURE 92

152. Spin the Rod

Let P be the raised center of the rod, let A be a point of attachment of a supporting rope to the ceiling, let C be the end of the raised rod that is connected to A, and let B be the point directly under and 24 feet below A. Because $\triangle ABC$ is a right triangle, with hypotenuse of length 25 and one leg of length 24, BC has length $\sqrt{625 - 576} = 7$. This means that $\triangle PBC$ is equilateral, so $\angle BPC = 60°$.

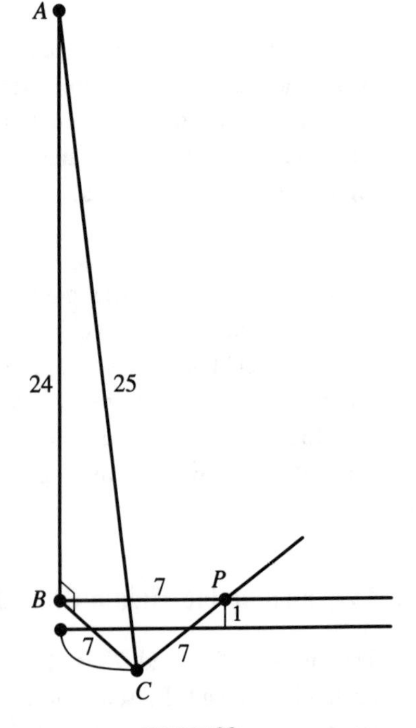

FIGURE 93

Problem 152.1 [Str, problem 106]. Suppose the length of the rod is a and the length of the rope by which it hangs is b. Suppose further that the twist (as above, about a vertical line through the rod's center) is through an angle θ. How much is the rod raised?

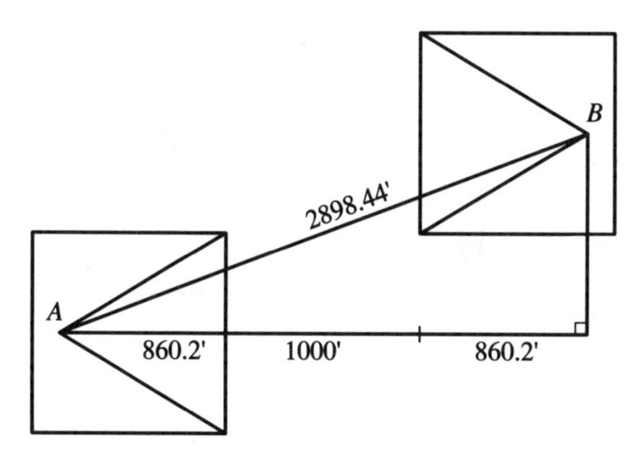

FIGURE 94

153. Traveling among the Pyramids

The shortest path is the one obtained by folding down the right-hand face of the leftmost pyramid and the left-hand face of the rightmost pyramid, thus reducing the problem to a 2-dimensional one in which the shortest distance is given by a straight line (see Figure 94). The altitudes of these triangular faces (from vertices A and B) have length $a = \sqrt{700^2 + 500^2} = 100\sqrt{74} = 860.23\ldots$ and so the final path is the hypotenuse of a right-angled triangle with legs $a + 1000 + a$ and 1000. Hence the shortest path has length $\sqrt{(2a + 1000)^2 + 1000^2} = 100\sqrt{496 + 40\sqrt{74}} = 2898.44\ldots$ feet. This argument justifies the minimum only for the set of paths that exit the leftmost pyramid on its right side and enter the other pyramid on the left side. But for any path that exits or enters on different sides one can find a path as just described whose length is shorter.

Note. This problem is due to Peter Riegel (Columbus, Ohio).

154. An Isosceles Tetrahedron

Imagine that faces PSR and PQR are made of wood, connected by a hinge along the edge PR, and edge QS is a rubber band. Pivot face PSR around the hinged edge PR until it is in the same plane as PQR, stretching the rubber band QS to a new length $d > c$. The result will be a parallelogram, with edge lengths a and b and diagonal lengths c and d. Since $d > c$, it is clear that $\angle PQR < 90°$. (This can be proven easily by the law of cosines.) Similar reasoning can be used to show that all face angles of the tetrahedron are acute.

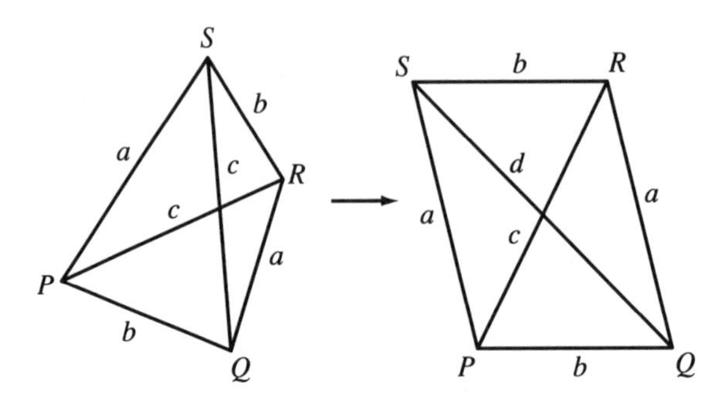

FIGURE 95

Note. This problem appeared on the first USA Mathematical Olympiad in 1972. Several solutions (all different from the one given here) can be found in [Kla2, 56–58].

155. Connect the Dots

There are precisely four points that are not in $L(L(V))$. It is helpful to inscribe the tetrahedron in a cube via six of the face diagonals (see Figure 96). Then the four unused corners of the cube are missing from $L(L(V))$.

To see this observe first that a line through points on adjacent (extended) tetrahedral edges lies in the plane of a tetrahedral face, and so misses the unused corners. And a line connecting one such corner to a point on a nearby edge lies in the plane of a face of the cube, and so misses the skew edge, which lies in the opposite plane.

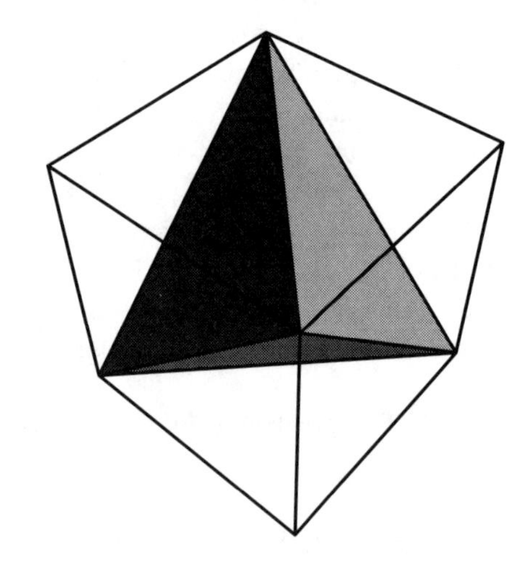

FIGURE 96

Problem 155.1. Show that the four points described in the solution above are the only points missing from $L(L(V))$. And show that if we start with an arbitrary tetrahedron with vertices A, B, C, and D, then the missing points are $A+B+C-D$, $A+B+D-C$, $A+C+D-B$ and $B+C+D-A$.

Notes. This problem is due to V. Klee [Kle2]. Let $A_n(X)$ be the n-dimensional affine hull of a point set X. Then Problem 155 shows that for 4 points in general position in 3-space, $A_4(X)$, which is all of $\mathbb{R}^3$, does not coincide with $A_2(A_2(X))$. For a discussion of general problems of this type involving affine and other sorts of hulls, see [Kle1].

156. A Point and a Plane

Let d be the distance from P, the fixed point, to the plane π. Then d is independent of the choice of the three perpendicular lines, so $1/d^2$ is also. Thus it is sufficient to prove $1/a^2+1/b^2+1/c^2 = 1/d^2$. To do this, choose a rectangular coordinate system with origin at P and such that positive x, y, z axes lie along the given segments. The equation for π is then $x/a + y/b + z/c = 1$. Recalling that $(1/a, 1/b, 1/c)$ is then a normal vector to the plane, we see that the perpendicular line from P to π has the parametric representation $t(1/a, 1/b, 1/c)$. This line intersects the plane at

$$t = t_0 = \frac{1}{\frac{1}{a^2} + \frac{1}{b^2} + \frac{1}{c^2}}.$$

So the distance from the origin to the plane is the distance from the origin to $t_0(1/a, 1/b, 1/c)$, which is $t_0\sqrt{1/a^2 + 1/b^2 + 1/c^2}$. Plugging in the value of t_0 gives $d = 1/\sqrt{1/a^2 + 1/b^2 + 1/c^2}$, as desired.

157. Painting a Cube

The first step is to compile the following table, assuming the large cube is $n \times n \times n$.

Number of faces painted	Number of small cubes with paint
1	n^2
2 (opposite)	$2n^2$
2 (adjacent)	$2n^2 - n$
3 (U-shaped)	$3n^2 - 2n$
3 (corner)	$3n^2 - 3n + 1$
4 (holes opposite)	$4n^2 - 4n$
4 (holes adjacent)	$4n^2 - 5n + 2$
5	$5n^2 - 8n + 4$
6	$6n^2 - 12n + 8$

Since 218, which equals $2 \cdot 109$, has none of the forms n^2, $2n^2$, $n(2n - 1)$, $n(3n - 2)$, $3k + 1$, or $4k$ the first six possibilities are eliminated. The next-to-last case is impossible because 218 is 2 (mod 4), while $5n^2 - 8n + 4$ is 0 or 1 (mod 4). Solving the quadratic equations that arise from the remaining two cases yields the solutions $n = 8$ and $n = 7$. So the two solutions are a $7 \times 7 \times 7$ cube with all faces painted and an $8 \times 8 \times 8$ cube with 4 faces painted. Since there is at least one unpainted face, the answer is $8 \times 8 \times 8$.

Note. This problem is based on one that George Berzsenyi posed for the 1983 Australian Mathematics Competition [Ber].

158. The Middle of a Moving Line

Let c be the perpendicular distance between the lines L and M and set up a coordinate system with origin at the midpoint of the shortest segment from line L to line M, and with x-axis above L and y-axis below M. Let $P = (a, 0, -c/2)$ and $Q = (0, b, c/2)$, and let d be the distance from P to Q. Then $a^2 + b^2 = d^2 - c^2$. But the midpoint of PQ has coordinates $(a/2, b/2, 0)$; it follows that the locus of the midpoint is the circle in the x-y plane given by $x^2 + y^2 = \frac{1}{4}(d^2 - c^2)$.

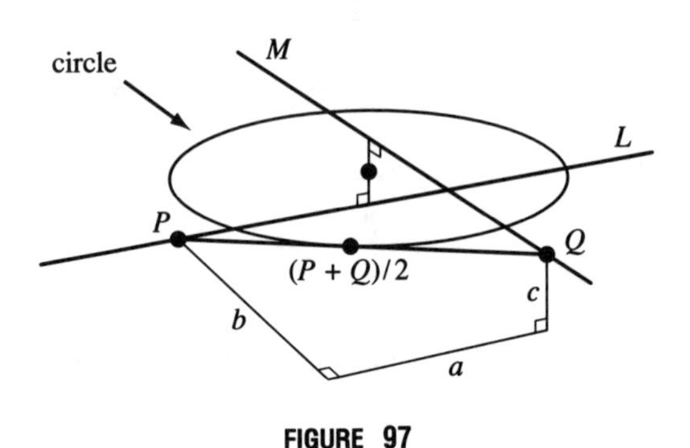

FIGURE 97

Note. This problem appears as problem 119 in [Str].

159. A Circle, a Sphere, and a Circle on a Sphere

It may come as a surprise that the two areas are equal! A simple case occurs when r, the radius of the circle in the plane, equals $\sqrt{2}R$, where R is the radius of the sphere. For then the compass cuts off an entire hemisphere, which has surface area $\frac{1}{2}4\pi(r/\sqrt{2})^2$, or πr^2, the area of the circle.

To solve the problem in general, we appeal to a famous, though equally surprising, theorem about spheres that was known to Archimedes. It might be called the crusty bread theorem, and will be useful to us in Problem 161 too. Imagine a spherical loaf of crusty bread, sliced into finitely many parallel slices, each having the same thickness. Then each slice contains the same amount of crust! An equivalent way of saying this is that the area of the crust on a slice is proportional to the thickness of the slice. Consequently, given a sphere of radius R (hence surface area $4\pi R^2$), a slice of width h has surface area $4\pi R^2\left[h/(2R)\right]$, or simply $2\pi Rh$. Before proving the crusty bread theorem, let us show why it solves Problem 159.

Let r be the radius of the circle in the plane and let R be the radius of the sphere. Then the circular region on the sphere may be viewed as the crust of an end-of-the-loaf slice with thickness h, and so its area is $2\pi Rh$. But, using Pythagoras twice (see Figure 98), $r^2 = h^2 + R^2 - (R-h)^2$, whence $r^2 = 2Rh$ and the area, $2\pi Rh$, is the desired πr^2.

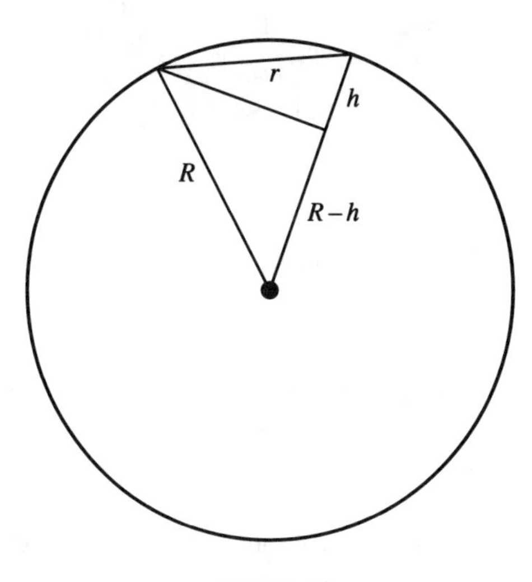

FIGURE 98

The crusty bread theorem is easily proved by calculus. We may assume the sphere has radius 1, and hence total area 4π. The sphere is the surface obtained by rotating the graph of $\sqrt{1-x^2}$ about the x-axis, and so we may use the surface area formula

$$\int_a^b 2\pi f(x)\sqrt{1+f'(x)^2}\,dx.$$

In this case the integrand reduces to simply 2π, and so the area is $2\pi(b-a)$, which is proportional to the thickness of the slice, as desired.

160. A New Dimension to a Famous Problem

It is easy to tile the right and front faces with $1 \times 1 \times 2$ dominoes, as illustrated. That leaves a box having dimensions $7 \times 7 \times 8$, and this too is easily tiled, by tiling each of the four $7 \times 7 \times 2$ sections using 49 tiles, all oriented vertically.

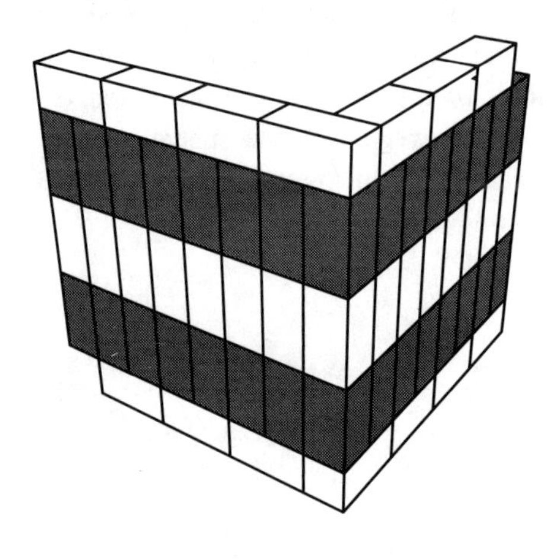

FIGURE 99

For no $n > 2$ does a tiling exist. We need only worry about $1 \times 1 \times 3$ and $1 \times 1 \times 5$ boxes since the number of cubes in C, 510, is not divisible by 4 or 7 or 8, and the $1 \times 1 \times 6$ case follows from the $1 \times 1 \times 3$ case. We present two proofs for $1 \times 1 \times 3$ tiles.

The first proof is a variation of the coloring argument for the chessboard case. We use the term *cubelet* for one of the $1 \times 1 \times 1$ cubes. The idea is to use three colors to color all the cubelets of C so that any tromino (a $1 \times 1 \times 3$ box) contains cubelets of all three colors. This can be done by labeling the cubelets using triples of integers between 0 and 7 [the deleted cubelets are $(0,0,0)$ and $(7,7,7)$], and then coloring the cubelet labelled (r,s,t) according to the value of $r+s+t$ modulo 3; we'll use red, green, and blue for 0, 1, and 2. Three cubelets in a row will necessarily have three consecutive integers as the values of $r+s+t$, so red, green, and blue are represented in each possible placement of a tromino.

Now, a count of the bottom plane of the unadulterated $8 \times 8 \times 8$ cube yields 21 red, 22 green, and 21 blue cubelets. Counting colors in the higher planes shows that the full cube has 170 red, 171 green, and 171 blue cubelets. But the deleted cubelets are red, so the adulterated cube has 168 red, 171 green, and 171 blue cubelets. These numbers would have to be the same if a tromino tiling of C existed.

Here is an alternate approach to the $1 \times 1 \times 3$ case. The bottom layer has 63 cubelets. This means that the number of vertical trominoes whose bottom cube is on the bottom layer is $0 \pmod 3$. The second layer has 64 cubelets. This means that the number of vertical trominoes whose bottom cube is in the second layer is $1 \pmod 3$. The third layer has 64 cubelets and so the number of vertical trominoes whose bottom cube is in this layer is $0 \pmod 3$, since we must count the ones coming up from the second layer. Continue in this manner to get that the number of vertical trominoes is $0+1+0+0+1+0+0+0 \equiv 2 \pmod 3$. Now rotate C and apply the same argument to get the same result for the number of horizontal trominoes in one direction. And rotate once more to again get $2 \pmod 3$ for the number of trominoes in the other horizontal direction. Thus the total number of trominoes would be $2+2+2 \equiv 0 \pmod 3$, contradicting the nondivisibility of 170 $(= 510/3)$ by 3.

Now, for the $1 \times 1 \times 5$ case observe first that the number of vertical tiles whose bottom cube is on the bottom layer is $3 \pmod 5$. The second layer has 64 cubelets, and so the number of vertical tiles whose bottom cube is in the second layer is $1 \pmod 5$, since we must count the $3 \pmod 5$ coming up from the bottom. Similarly, the number of vertical tiles whose bottom cube is in the third layer is $0 \pmod 5$ (from $64-3-1$). And therefore the same is true for vertical tiles whose bottom cube is in the fourth layer. But this means that there are no vertical tiles whose top cube is in the top layer, a contradiction, since the number of such must be $3 \pmod 5$.

Problem 160.1. Assume the $8 \times 8 \times 8$ cube is colored in chessboard fashion. Show that if a white cubelet and a black cubelet are removed, then the mutilated $8 \times 8 \times 8$ cube can be tiled with $1 \times 1 \times 2$ boxes.

Note. Problem 160, the alternate solution, and problem 160.1 are due to Moshe Rosenfeld (Pacific Lutheran University).

161. Surveying on Earth

The first thing to observe is that Alice and Bob do not live in Canada or Alaska (thus nearby bears are not white). For the only way their land could satisfy the conditions is if the plots straddled the equator. To see this, imagine a "square" piece of land in the northern hemisphere. If the north–south (N–S) fences were truly N–S, then the fence forming the top boundary would be smaller, if only by a little bit, than the fence forming the bottom boundary (see Figure 100), contradicting the assumption. In fact, it must be the case that the equator exactly bisects each piece of land. The diagram shows, with exaggerated dimensions, what the plots of land look like. The grid is drawn so that the latitude at the north end of

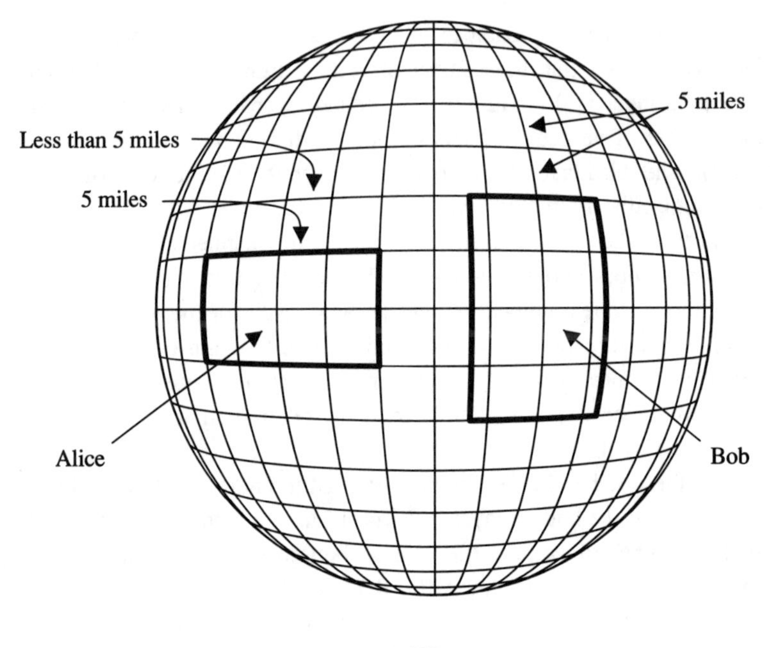

FIGURE 100

Alice's land is divided into 5-mile segments. This means that the segments at the equator are longer than 5 miles and the segments at Bob's north end are less than 5 miles.

Now, one might think about the two pieces and conclude that Bob's has the larger area since the width of the middle of Bob's land is more than half the width of Alice's, while his eastern and western boundaries are exactly twice the length of Alice's. To verify that this heuristic is correct, we take a more technical approach that computes the areas exactly. Although calculus can be brought to bear, we avoid a direct use of calculus by appealing to the crusty bread theorem, mentioned in the solution of Problem 159. That theorem, which follows from elementary calculus, told us that on a sphere of radius R, a slice of width h has surface area $2\pi Rh$. This method of attack was pointed out to us by G. C. Shephard (East Anglia University, England).

Let θ be the latitude of Bob's northern boundary, measured in radians (i.e., the angle subtended at the center of the earth by the upper half of Bob's N–S fence). The *entire* region between latitude θ and $-\theta$ has area $2\pi Rh$ where h is the vertical thickness of the slice; but h is $2R\sin\theta$ so this slice of earth has area $4\pi R^2 \sin\theta$. It remains to figure out what proportion of the slice is occupied by Bob's land. His land covers an arc 10 miles long at its northerly border, compared to the total arc at that latitude of $2\pi R\sin(\pi/2 - \theta)$, or $2\pi R\cos\theta$. Therefore Bob's land has area

$$4\pi R^2 \sin\theta \frac{10}{2\pi R\cos\theta} = 20R\tan\theta.$$

Recalling the circular arc formula $s = R\theta$, and observing that the upper half of Bob's N–S fences are 10 miles long, we see that $10 = R\theta$, so $\theta = 10/R$ and the area of Bob's land is $20R\tan(10/R)$. Similarly, Alice's area is $40R\tan(5/R)$. Since $\tan 2\theta > 2\tan\theta$ for $0 < \theta < \pi/4$, Bob's land is larger. Using $R = 3950$, we find that Bob's area is 200.000427... square miles and Alice's is 200.000106... square miles. So Bob has about 992 square yards more than Alice, or about $\frac{1}{6000}$ of 1 percent more land.

CHAPTER 12

Miscellaneous

162. A Puzzling Reflexicon

One starts with the five-letter vertical slot, which must be "ONE ?" Then the horizontal slot containing the question mark cannot be a 2 or 10, for there would then not be one "O" or "N", respectively. So it must be a 6. The slot containing the "S" in "SIX" must be 13, 14, or 18. It can't be 14 because F begins no 5-letter number. The number of Ss cannot be 18 because there can't be enough 7s to make up the excess S's. Therefore the long vertical slot is "THIRTEEN SS"; the slot at its top must then be a 3. The "R" in "THREE" must belong to a 4.

The "THREE ?S" can't have an N for that would make 3 Ns and we know there is one more from the missing "SEVEN", which must exist to make the Ss come out right. So the vertical slot containing the THREE's question mark must be 4 or 5. This means that the 7 is not in the horizontal slot (too many Es), so it must be in the vertical slot at the far right. We now know that the twelve letters are EFHINORSTUVX. There is no G, so no 8s. The slot with the V of 7 is therefore a 3. Taking inventory of the letters used and continuing in this vein leads to the unique solution illustrated.

Note. This puzzle is due to Lee Sallows [Sal1].

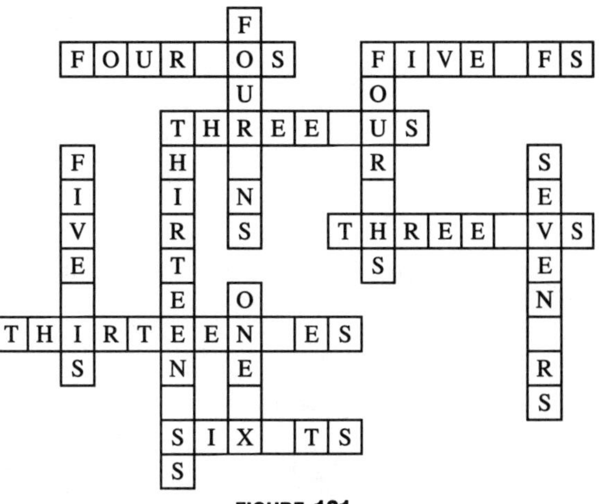

FIGURE 101

163. A Complicated Constant

Let $f(x) = \sqrt{x + 2\sqrt{x-1}} + \sqrt{x - 2\sqrt{x-1}}$. Then $f(x)$ has the constant value 2 for $1 \le x \le 2$. To see why, first rewrite $f(x)$ as follows:

$$\begin{aligned} f(x) &= \sqrt{(x-1) + 2\sqrt{x-1} + 1} + \sqrt{(x-1) - 2\sqrt{x-1} + 1} \\ &= \sqrt{(\sqrt{x-1}+1)^2} + \sqrt{(\sqrt{x-1}-1)^2} \\ &= \sqrt{x-1} + 1 + \left|\sqrt{x-1} - 1\right|, \end{aligned}$$

where the absolute value signs are needed because the radical sign means nonnegative square root. It is now clear that

$$f(x) = \begin{cases} \text{undefined} & \text{if } x < 1 \\ 2 & \text{if } 1 \le x \le 2 \\ 2\sqrt{x-1} & \text{if } x > 2. \end{cases}$$

The graph of f is shown in the diagram.

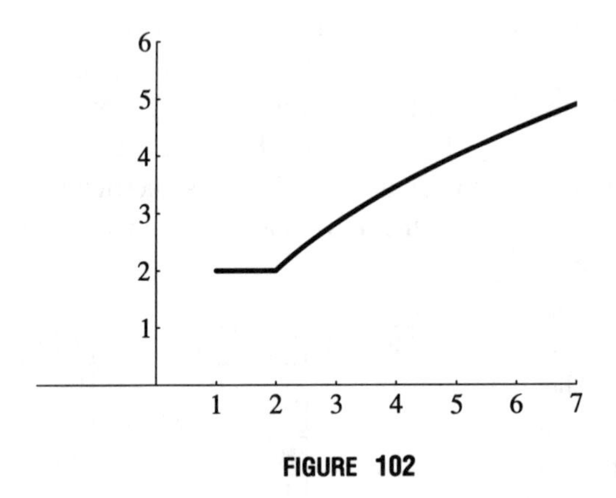

FIGURE 102

Note. This problem is discussed in [AB], along with other similarly counterintuitive problems.

164. Conservation of Blits

Figure 103 illustrates the best-known general method ($4n - 3$ blits to transpose an $n \times n$ array):

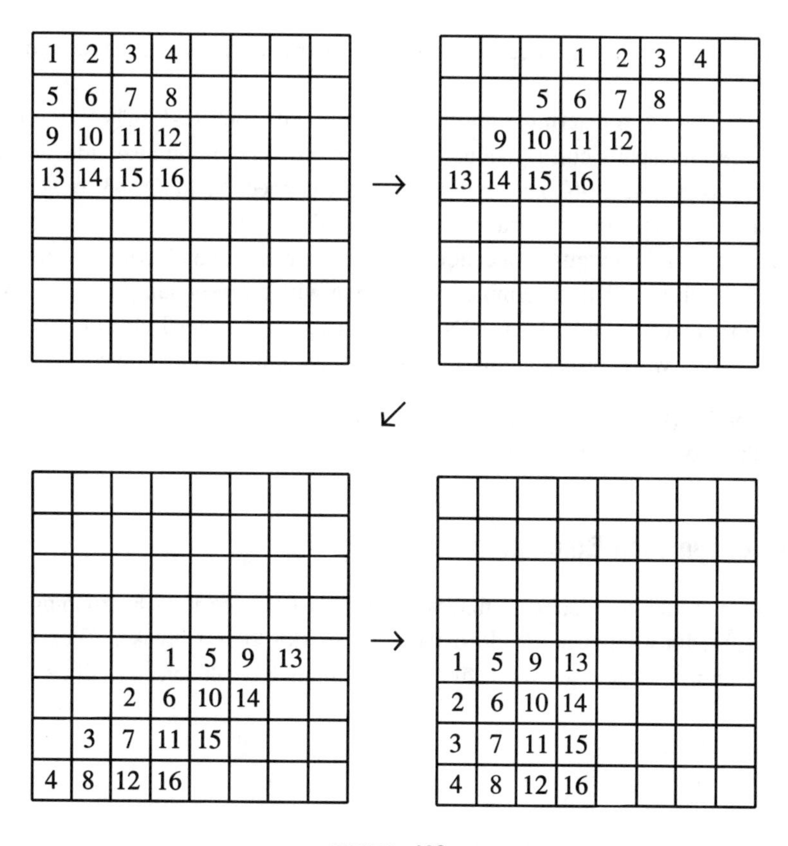

FIGURE 103

The first and last transformations each use $n - 1$ blits; the other transformation uses $2n - 1$ blits. Thus the total is $4n - 3$, or 97 for the 25×25 case. It is not known whether this is best possible.

Note. This problem, which has its origin in computer graphics, was suggested by Jim Guilford (Digital Equipment Corporation, Hudson, Mass.)

Problem 164.1. With the same setup as in Problem 164, how many blits do you need to rotate a matrix 90°?

165. The Electrician's Dilemma

Here is a method that shows how to handle the case of an odd number of wires with just two flights. Begin by joining the wires in pairs, leaving one isolated; label the latter #1, and call the wires that are paired partners. Fly to the other end of the tunnel.

Use electricity to determine the isolated wire and label it #1. Check the others, group them into pairs corresponding to pairs of partners on the near side, and label the pairs as (#2, #3), (#4, #5), and so on. Now connect #1 to #2, #3 to #4, and so on, leaving #n isolated. Fly back to the near side.

Disconnect all previous connections (but keep track of which wires are partners). Feed electricity to #1 to determine which is the wire corresponding to #2 on the far side; label that wire #2. Then #2's partner must be #3. Feed electricity to #3 to determine the wire corresponding to #4. Continue until all is resolved.

Problem 165.1. What is the minimum number of helicopter flights necessary to solve the problem for n wires, where n is even?

166. An Egg-Drop Experiment

The diagram shows a scheme that will work in no more than 8 droppings. The thick (thin) lines tell what to do if the egg breaks (doesn't break). Thus we start at the eighth floor and go to the first floor if the egg breaks and the fifteenth floor

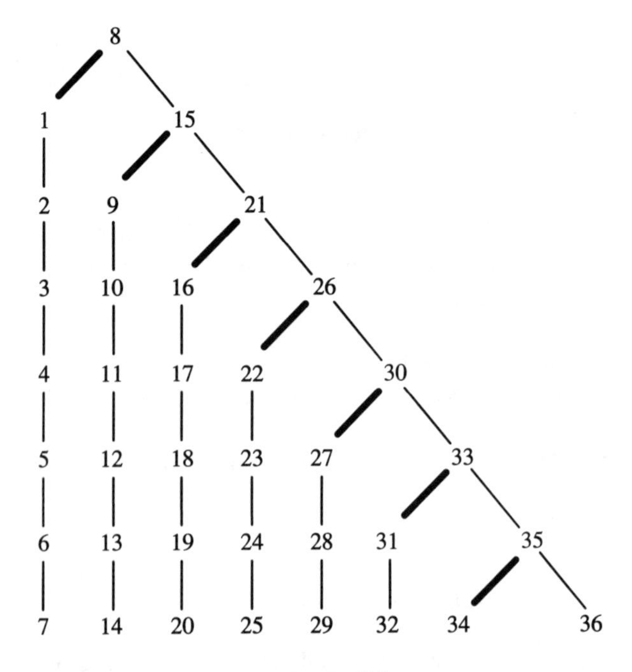

FIGURE 104

if it does not, and so on. If there were a 7-dropping solution, then the first drop must be from the seventh floor, the second drop could be from no higher than the thirteenth floor, and so on; this would not reach the top floor.

Problem 166.1. Solve the problem for an n-story building, still with two eggs. How about an n-story building with k eggs?

167. The Race Goes to the Swiftest

It is a little surprising that Bob can indeed meet the conditions of the problem and still win the race. Partition the distance into alternating segments of 0.2 miles and 0.8 miles. There will be 27 short segments and 26 long segments. The idea is that Bob might run at constant speed on each of these segments, a little faster than average on the short segments and slower than average on the long segments. This will mean that each mile on the course will be covered in the same time. Precisely, let x and y be the times, in seconds, taken to run the short and long segments, respectively. Then every mile interval will take $x+y$ seconds. Now, the conditions are:

$$\begin{aligned} x+y &= 481 && \text{(8 minutes and 1 second)} \\ 27x+26y &= 209.6\times 60-1 && \text{(Alice's time, minus one second).} \end{aligned}$$

This is easily solved: $x = 69$ and $y = 412$. So, by running the 0.2-mile segments in 69 seconds and the 0.8-mile segments in 412 seconds, Bob will lose infinitely many one-mile battles, but will win the 26.2-mile war.

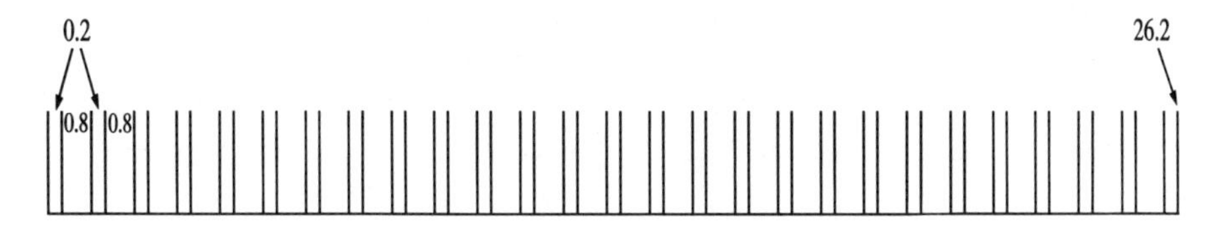

FIGURE 105

Note. This problem first appeared in [Wag1].

168. A Competition Conundrum

The total number of points for the entire competition was $22+9+9=40$, so we have $n(p_1+p_2+p_3)=40$, where n is the number of events. Clearly $p_1+p_2+p_3 \geq 3+2+1=6$, so the only possible values for n are 1, 2, 4, and 5.

Since Bob won the 100-yard dash and his total score was 9 points, we know that $p_1 \leq 9$. But then Alice could not have achieved a score of 22 points unless there were more than 2 events. Therefore n is either 4 or 5.

Suppose first that $n = 4$, so $p_1 + p_2 + p_3 = 10$. Since Alice earned 22 points in 4 events, we must have $p_1 \geq 6$. But then the only way that Bob could win the 100-yard dash and end up with only 9 points is if he came in third in the other 3 events, $p_1 = 6$, and $p_3 = 1$. Therefore $p_2 = 3$. But now even if Alice came in second in the 100-yard dash and won every other event, her score would only be $3 \cdot 6 + 3 = 21$, contrary to the given information.

Therefore $n = 5$, and $p_1 + p_2 + p_3 = 8$. As before, since Alice earned 22 points in 5 events we must have $p_1 \geq 5$, and since Bob won the 100-yard dash and only earned 9 points he must have come in third in all of the other events, and we must have $p_1 = 5$ and $p_3 = 1$. Therefore $p_2 = 2$. The only way that Alice could have received 22 points is if she came in second in the 100-yard dash and won every other event. Therefore Eve must have come in third in the 100-yard dash and second in every other event, including the high jump.

169. A Messy Desk

Imagine that the desktop was painted just before the paper was put on it. Then some of the sheets of paper will have picked up some paint when they were put on the desktop. More precisely, each sheet of paper will have picked up paint on it from any part of the desk that it covers that is not covered by any sheet lower in the pile. Since the desktop is completely covered by the paper, the total area of paint on all the sheets of paper is exactly equal to the area of the desktop. If we remove the five sheets with the least paint on them, then clearly the total area of paint on the sheets remaining is at least two-thirds of the area of the desktop, and therefore at least two-thirds of the desktop is covered by the remaining sheets.

170. An Optimist and a Pessimist

Let d_n be the total distance traveled by Bob after n steps, and s_j the size of his jth step. Then $d_n = \sum_{j=1}^n s_j$ and $s_j = u_j(1 - d_{j-1})$. Note that d_n is increasing and therefore, for $1 \leq j \leq n$, $s_j > u_j(1 - d_n)$. Therefore $d_n > \sum_{j=1}^n u_j(1 - d_n)$, so

$$1 - d_n < \frac{d_n}{\sum_{j=1}^n u_j} < \frac{1}{\sum_{j=1}^n u_j}.$$

Now since the infinite sum of the u_js diverges, this last term goes to 0, so d_n converges to 1.

Problem 170.1. Show that if Alice cannot travel arbitrarily far, then Bob cannot get arbitrarily close to the chair.

171. The Ambiguous Clock

Take two copies of a standard clock: one that runs at normal speed and an auxiliary clock whose motion is determined by the physical attachment of its hour hand to the minute hand of the normal clock. This will cause the auxiliary clock to run 12 times as fast as the normal one. Now, the hour hand of the auxiliary clock is forcibly aligned with the minute hand of the other; thus an ambiguous setting is obtained whenever the minute hand of the auxiliary clock is aligned with the hour hand of the normal clock (and all four hands are not in the same position). But the auxiliary minute hand goes 144 times as fast as the normal hour hand. Thus it makes 144 revolutions while the normal hour hand makes just 1. This means that the auxiliary hour hand is aligned with the normal hour hand 143 times in a 12-hour period. However, at 11 of these times all four hands will be aligned, since the minute and hour hands of a single clock are aligned with each other 11 times in a 12-hour period. At these times, the reading is not ambiguous as far as the time of day is concerned (though there is ambiguity as to which hand is which). So the answer is 143 − 11, or 132.

In fact, the 143 alignment times are given by $n\left(\frac{12}{143}\right)$ hours after noon, $n = 0, 1, 2, \ldots, 142$. The unambiguous times (hour/minute alignment) are the entries in this sequence corresponding to $n = 0, 13, 26, \ldots, 130$.

Problem 171.1 (Gerald Weinstein [Wei]). Imagine a clock with three hands of the same length: an hour hand, a minute hand, and a second hand. Can one always tell the time?

Notes. The use of an auxiliary clock yields a solution that is more elegant than a straightforward application of algebra. Chris Coyne, an undergraduate at the University of Minnesota, came up with this approach, though it had also been discovered by K. A. Post [Pos]. This problem has surfaced many times, from Dudeney's classic puzzle book [Dud, problem 48], to later journals [Cam, Pie, McC, Szi, Wei].

172. A Problemist's Joke

Let b be Alice's base. Note that $b = 10_b$, but b is not necessarily the number ten. Now, Alice's statement tells us that the two bases are b and $b+4$. From Bob's "25", a number that cannot be less than Alice's "26", we know that Bob is using the larger base, $b+4$. Both sentences together yield that $25_{b+4} = 1+26_b+22_b-13_{b+4}$. This means that $2b+13 = 1+2b+6+2b+2-b-4-3$, or $b = 11$. Therefore Binumeria has 25_{15}, or 35 residents.

Note. This problem appeared in [Rab2].

173. Throw Your Rings into the Hat

There are three solutions; the first one below is perhaps the easiest to find.

Hat A	Hat B	Hat C	Hat D	Hat E
1	2	3	4	5
8	9	10	6	7
15	11	12	13	14
17	18	19	20	16
24	25	21	22	23

Hat A	Hat B	Hat C	Hat D	Hat E
1	2	12	11	4
3	6	14	13	8
5	15	16	17	10
7	20	23	24	18
9	22			25
19				
21				

Hat A	Hat B	Hat C	Hat D	Hat E
1	2	11	12	8
3	4	13	14	15
5	6	17	16	20
7	10	24	23	22
9	18			
19	25			
21				

Notes. The problem arose from some geometrical investigations of Lee Sallows [Sal2]. Suppose we seek a certain type of walk that uses straight steps in the plane and:

- starts at the origin;
- has a first step of length 1;
- has subsequent steps of length 2, 3, 4, and so on;
- turns right 72° or left 72° at the end of each step;

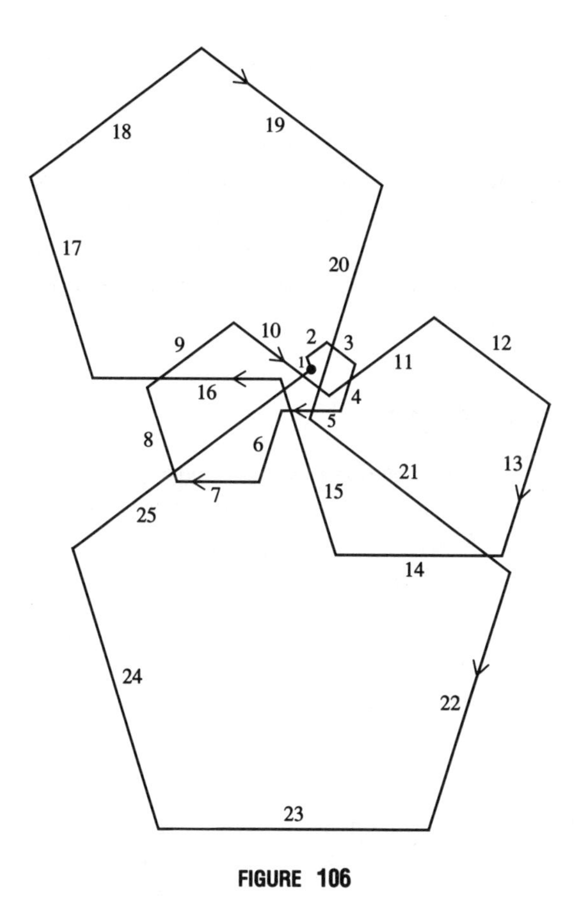

FIGURE 106

- finishes back at the origin in such a way that a 72° turn after the last segment yields the direction of the first segment.

Such a walk is called a *serial isogon.* Then a solution to the hat problem leads to a serial isogon, because a hat solution may be interpreted as turning instructions for constructing the isogon, where each hat corresponds to one of the five possible directions of motion. For example, the first solution above corresponds to the turn-sequence RRRRLRRRRLRRRRLRRRRLRRRRL, where R and L denote right and left, respectively (Figure 106 shows the actual walk). The walk returns to the origin because its total effect is the same as that of a walk around a regular pentagon of side-length 65. The other two hat-solutions lead to two other serial isogons.

Problem 173.1. Find a serial isogon that uses 90° turns instead of 72° turns.

Problem 173.2. Prove that in any serial isogon with 90° turns, the number of steps is a multiple of 8.

The hat problem is due to Lee Sallows and Hans Cornet (Nijmegen, The Netherlands); see [Sal2]. Problems 173.1 and 173.2 are discussed in [SGGK]. While the argument above shows that a solution to the hat problem leads to a serial isogon, the converse is not necessarily true. However, in the case of a turning angle of 72° there are indeed only three serial isogons [Sal2, p. 63], and therefore only three solutions to the hat problem.

174. A Super Exponential Integral

Because $x^x = e^{x \ln x}$, the substitution $u = x \ln x$ looks promising, but to use it we must also substitute $du = (1 + \ln x)\, dx$. Introducing a factor of $1 + \ln x$ into the integral leads us to the following lemma.

Lemma. *If* $1 \le a < b$ *then*

$$\frac{b^b - a^a}{1 + \ln b} \le \int_a^b x^x\, dx \le \frac{b^b - a^a}{1 + \ln a}.$$

Proof. For $a \le x \le b$,

$$x^x = \frac{e^{x \ln x}(1 + \ln x)}{1 + \ln x} \le \frac{e^{x \ln x}(1 + \ln x)}{1 + \ln a}.$$

Thus

$$\int_a^b x^x\, dx \le \int_a^b \frac{e^{x \ln x}(1 + \ln x)}{1 + \ln a}\, dx = \int_{a \ln a}^{b \ln b} \frac{e^u\, du}{1 + \ln a} = \frac{b^b - a^a}{1 + lna}.$$

A similar calculation leads to the claimed lower bound for the integral.

We now apply the lemma to the given integral, getting

$$1.78407 \cdot 10^{199} \approx \frac{100^{100} - 1}{1 + \ln 100} \le \int_1^{100} x^x\, dx \le 100^{100} - 1 \approx 10^{200}.$$

Unfortunately, this doesn't give us sufficient accuracy.

Because x^x grows very quickly, we concentrate on the right end of the interval. Splitting the integral at 99 and applying the lemma to both halves, we get

$$\frac{99^{99} - 1}{1 + \ln 99} + \frac{100^{100} - 99^{99}}{1 + \ln 100} \le \int_1^{99} x^x\, dx + \int_{99}^{100} x^x\, dx \le 99^{99} - 1 + \frac{100^{100} - 99^{99}}{1 + \ln 99},$$

or

$$1.78408 \cdot 10^{199} \leq \int_1^{100} x^x \, dx \leq 1.81764 \cdot 10^{199}.$$

Taking the average of these upper and lower bounds gives us an answer of $1.80086 \cdot 10^{199}$, which is in error by at most half the difference, or $1.678 \cdot 10^{197}$. This is just under 1% of the lower bound for the integral.

Using *Mathematica*'s numerical integrator yields:

```
NIntegrate[x^x, {x, 1, 100}]
1.78464 10^199
```

Our estimate of 1.80086 differs from this value by a little less than 1%.

Note. This problem is due to Murray Klamkin [Kla1, pp. 134, 159].

175. Passion Has No Square Root

First, KISS $< \sqrt{10,000,000} \approx 3162\ldots$, so K is one of 1, 2, 3 and if K is 3 then $\text{I} \leq 1$. And S must be one of 2, 3, 4, 7, 9 since PASSION does not end in S (8 is eliminated because $88 \cdot 88 = 7744$). Because SS^2 ends in ON and ISS^2 ends in ION, we get the following chart of possibilities.

S	ON	I	K	KISS
2	84	–		
3	89	0	1 or 2	1033 2033
		2	1 or 2	1233 2233
		4	1 or 2	1433 2433
		6	1 or 2	1633 2633
4	36	–		
7	29	–		
9	01	6	1 or 2	1699 2699

Of the choices for KISS, only 2033 squares to the right sort of PASSION, 4133089.

Note. This puzzle is due to Alan Wayne [Way], who also invented the classic alphanumeric puzzle: TEN + TEN + FORTY = SIXTY.

176. A Problem Fit for a King

The string ABCADBECDE has no template. The reason is that B and C are adjacent to each other and to A, D, and E, and A is adjacent to D is adjacent to

E. But the squares that are simultaneously neighbors of any two adjacent squares (B and C) comprise either two adjacent squares (if B and C meet at a corner) or a disjoint pair of two adjacent squares (B and C meet on a side). In neither case do the simultaneous neighbors contain three squares, with one adjacent to the two others.

Notes. In graph theory terms, the string ABCADBECDE corresponds to an Eulerian path of length 9 in the graph K_5 minus an edge (the vertices are A, B, C, D, and E), and this graph is not a subgraph of the lattice adjacency graph (vertices are squares and two squares are connected if they are a king's move apart).

This problem is due to Lee Sallows [Sal3], who has shown that there are exactly 22 distinct 10-strings that arise from K_5 minus an edge:

1.	1231425345	8.	1234135425	15.	1234254135
2.	1231425435	9.	1234152453	16.	1234254153
3.	1231435245	10.	1234154253	17.	1234513524
4.	1231435425	11.	1234251354	18.	1234514253
5.	1231452435	12.	1234251453	19.	1234524135
6.	1231453425	13.	1234253145	20.	1234524153
7.	1234135245	14.	1234253154	21.	1234531425
				22.	1234531524

None of these patterns is realized by an English word, however. The closest, for mnemonic purposes, is the string "Insciences," which corresponds to number 9 in the list.

The authors have shown, by a proof involving some tedious enumeration of possibilities, that every graph with 9 or fewer edges embeds in the lattice, with three exceptions: K_5 minus an edge, $K_{3,3}$, and $K_{1,9}$. It follows that every 9-string has a template and, because $K_{3,3}$ and $K_{1,9}$ are not Eulerian, that the 22 examples of Sallows are a complete list of the 10-strings that fail to have a template.

177. Weigh The Boxes

Let the weights of the boxes, in increasing order, be a, b, c, d, and e. Then the two smallest weights of pairs of boxes are $a+b$ and $a+c$, so we have

$$a + b = 110 \qquad (1)$$

$$a + c = 112 \qquad (2)$$

Similarly,

$$d + e = 121 \tag{3}$$

$$c + e = 120 \tag{4}$$

Since each box is used in four weighings,

$$4(a+b+c+d+e) = 110+112+113+114+115+116+117+118+120+121,$$

so

$$a + b + c + d + e = 289 \tag{5}$$

Adding (1) and (3) we get $a + b + d + e = 231$, and combining this with (5) gives us $c = 58$. Then from (2) and (4) we get $a = 54$ and $e = 62$, and finally from (1) and (3) we get $b = 56$ and $d = 59$. So the weights are 54, 56, 58, 59, and 62 pounds.

178. A Question of Imbalance

The minimum number of uses of the balance is seven. To rank-order the coins in seven weighings, first weigh A against B and then C against D. By relabeling the coins if necessary, we may assume without loss of generality that A is lighter than B and C is lighter than D. Now weigh A against C, and assume, again without loss of generality, that A is lighter than C. If we let each letter stand for the weight of the corresponding coin, then we now know $A < C < D$ and also $A < B$, and we have used three weighings. We can now determine the ranking of E relative to A, C, and D in two more weighings, by weighing E first against C, and then against A if E is lighter than C, and D if it is heavier. Finally, since we know $A < B$ all that remains is to determine the ranking of B relative to C, D, and E, and this can also be done in two weighings.

To see that seven weighings are required, note first that in n weighings we can only distinguish 2^n possible orderings of the coins, since each weighing has two possible outcomes. (We may ignore the possibility that the two sides in a weighing balance exactly, because the coins can be chosen so that no two collections of coins weigh the same amount.) But the number of possible orderings of the coins is $5!$, or 120, so the rank-ordering cannot be accomplished in n weighings unless $2^n \geq 120$. Since $2^6 = 64$, seven weighings are required.

Note. This problem appeared in [Ste2, problem 52].

179. Catch the Counterfeit

Let the coins be called A, B, C, D, E, and F. Let x be the weight of a genuine coin, and let $x+\delta$ be the weight of the counterfeit coin, if there is one, where δ could be positive or negative but $\delta \neq 0$. First weigh A, B, and C together, and then weigh B, C, D, and E together. If A, B, C, D, and E are genuine then the ratio of these weights will be $\frac{3}{4}$, but if one of them is counterfeit then the ratio will be either $(3x+\delta)/4$, $3x/(4x+\delta)$, or $(3x+\delta)/(4x+\delta)$, none of which is equal to $\frac{3}{4}$. Thus, if the ratio is $\frac{3}{4}$ then all of these coins are genuine, and they all weigh one third of the result of the first weighing. To complete the process, simply weigh coin F on the third weighing.

Now suppose the ratio of the first two weights is not $\frac{3}{4}$. Then one of A, B, C, D, and E must be counterfeit. On the third weighing, weigh C and D. The following table shows what the results of the three weighings would be for each possible choice of the counterfeit coin:

		Counterfeit coin				
		A	B	C	D	E
	1	$3x+\delta$	$3x+\delta$	$3x+\delta$	$3x$	$3x$
Weighing number	2	$4x$	$4x+\delta$	$4x+\delta$	$4x+\delta$	$4x+\delta$
	3	$2x$	$2x$	$2x+\delta$	$2x+\delta$	$2x$

Let the results of the three weighings be w_1, w_2, and w_3. Note that if A is the counterfeit coin then $w_2/w_3 = 2$, and this is not true in any other case. Thus, if $w_2/w_3 = 2$ then A is counterfeit, the weight of all the coins except A is $w_3/2$, and the weight of A is $w_1 - w_3$. Similarly, each of the other cases can also be distinguished by a particular algebraic relationship among w_1, w_2, and w_3 that holds in that case and no others, and in each case once the counterfeit coin is identified it is easy to compute the weights of all the coins. The relationships are:

If B is counterfeit: $w_3 = 2(w_2 - w_1)$.
If C is counterfeit: $w_2 - w_1 = w_1 - w_3$.
If D is counterfeit: $w_1 = \frac{3}{2}(w_2 - w_3)$.
If E is counterfeit: $w_1/w_3 = \frac{3}{2}$.

For Further Investigation. In our solution above we have assumed that the scale is perfectly accurate. What if we make the more realistic assumption that there is some small but positive bound ε on the error of the scale? For what values of x and δ would you be able to determine all the weights in three weighings?

Note. A similar problem appeared in [Mau].

180. And Now For Something Completely Different

	Queen	King	
1.	a2 check	b4	(if Kc3 then: Qa4, Kb2, Qb4 check, continue as from 4)
2.	c2	a3	
3.	c4	b2	
4.	b4 check	c1	(the other king moves, a2 or c2 or a1, are easier for white)
5.	c3 check	b1	
6.	b3 check	c1	
7.	b4	c2	
8.	a3	b1	
9.	c3	a2	
10.	b4	a1	(not all moves are forced; these are black's best plays)

Notes. This problem was originally posed by Martin Gardner (see [GL]). A generalization of this result that states that the queen can force the king to her starting corner on an $n \times m$ board if and only if $n \neq m$ is proved in [CLL]. In the generalization to a square board, the king moves first, for otherwise the position would be illegal.

181. Guess My Numbers

Bob can always determine Alice's numbers in only two rounds. In the first round, Bob chooses $a_1 = a_2 = \cdots = a_n = 1$, and Alice tells him the number $S = x_1 + x_2 + \cdots + x_n$. Note that, since all Alice's numbers are positive, for all i, $x_i \leq S$. On the next round, Bob chooses $b_1 = 1$, $b_2 = S + 1$, $b_3 = (S+1)^2, \ldots, b_n = (S+1)^{n-1}$, and Alice tells him the number

$$N = x_1 + x_2(S+1) + x_3(S+1)^2 + \cdots + x_n(S+1)^{n-1}.$$

Now, to determine Alice's numbers Bob simply writes N in base $S+1$; the digits of N in base $S+1$ will be Alice's numbers.

Bob cannot, in general, determine Alice's numbers in only one round. For example, suppose Alice's response to Bob's first question is $a_1 + a_2 + \cdots + a_n + a_1 a_2$. Then Bob can't tell whether Alice's numbers are

$$x_1 = a_2 + 1,\ x_2 = x_3 = \cdots = x_n = 1 \quad \text{or}$$
$$x_1 = 1,\ x_2 = a_1 + 1,\ x_3 = x_4 = \cdots = x_n = 1.$$

182. Divide and Conquer

Alice's move will leave Bob with an even number of pennies in one of the boxes and an odd number in the other. He can discard the odd one and split the even number into 1 and an odd number. Alice must then discard the singleton, again presenting Bob with an opposite-parity situation. Eventually Bob will see 2 pennies in one of the boxes. He discards the other box, splits the 2 into $1 + 1$, and wins. This provides a completely general analysis: if the initial position consists of two odd numbers, Alice loses; if the initial position includes one even number, Alice wins.

Note. This problem is due to Keith Austin and appeared in [Stew, chapter 11].

183. Divide and Be Conquered

Call an integer that is greater than 1 and congruent to either 0 or 1 $(\text{mod } 3)$ a *winning number*, and the numbers greater than 1 that are congruent to 2 $(\text{mod } 3)$, *losing numbers*. The winning strategy is to always leave your opponent with either two losing numbers, or a 1 and a losing number. This can always be done by a player facing at least one winning number. For if the winning number is n and congruent to 0 $(\text{mod } 3)$, then he or she leaves 1 and $n - 1$. If n is congruent to 1 $(\text{mod } 3)$, then the pair 2 and $n - 2$ is left. The opponent will have to split a losing number and so will always leave either two 1s (which is a losing position in the original game, and so a winning position here) or at least one winning number, which allows the strategy to be continued. Since 51 is divisible by 3, Alice discards 101, and splits 51 into 1 and 50, confident that she will eventually lose, and therefore win.

184. Greed vs. Ethics in Gambling

Yes, Alice should agree to Bob's terms. For once she knows Bob's choices, she can make selections that are the exact opposite of Bob's; this puts Eve at a big disadvantage. For suppose Bob picks $(0, 0, 0)$ and Alice picks $(1, 1, 1)$, where 0 (resp., 1) denotes a prediction of a home-team loss (resp., win).

If Eve's choice is $(0, 0, 0)$, her expected return is $\frac{1}{2}(\$15.00) + \frac{1}{2}(\$0)$ or \$7.50. If Eve's choice is $(1, 0, 0)$, $(0, 1, 0)$, or $(0, 0, 1)$, her expected return is $\frac{1}{8}(\$30.00) + \frac{2}{8}(\$15.00) + \frac{5}{8}(\$0)$, or \$7.50. Similar reasoning shows that her expected return is \$7.50 no matter what choices she made. Alice's and Bob's expected joint return is therefore \$22.50, so, because of symmetry, Alice's expected return is \$11.25, enough to overcome the \$1 premium paid to Bob. Bob really makes out like a bandit, however, since his expected profit on this deal is \$2.25 (\$11.25 + \$1 − \$10.00).

Notes. It turns out that in pools of this type it is always advantageous for two players to team up and make diametrically opposed choices. For more details see [DDS], which contains a comprehensive analysis. The most profitable situation is with 5 players and 1 game; then collusion between two players leads to an expected return of \$13.40 for a \$10.00 bet.

185. A Trigonometric Surprise

If f_n denotes the nth Fibonacci number ($f_0 = 0$, $f_1 = 1$, $f_2 = 1$, $f_n = f_{n-1} + f_{n-2}$), then $a_n = \sqrt{f_n/f_{n+1}}$. To prove this, first note that each a_n is positive (this uses $0 < \arctan x < \pi/2$ for $x > 0$). Then simple trig yields

$$a_{n+1} = \frac{1}{\sqrt{1+a_n^2}},$$

whence $a_1 = 1$, $a_2 = 1/\sqrt{2}$, and $a_3 = \sqrt{2/3}$. Now, assume inductively that $a_k = \sqrt{f_k/f_{k+1}}$. Then

$$a_{k+1} = \frac{1}{\sqrt{1+a_k^2}} = \frac{1}{\sqrt{1+\frac{f_k}{f_{k+1}}}} = \sqrt{\frac{f_{k+1}}{f_k + f_{k+1}}} = \sqrt{\frac{f_{k+1}}{f_{k+2}}}.$$

Because the limit of f_{n+1}/f_n is the golden ratio—$\frac{1}{2}(1+\sqrt{5})$—it follows that $\lim_{n\to\infty} a_n$ is the reciprocal of the square root of the golden ratio.

186. Complex, Yet Simple

(a) Because $b = -a$, this is immediate.
(b) Interpret the complex numbers as vectors $\mathbf{u}$, $\mathbf{v}$, $\mathbf{w}$ emanating from the origin, and let θ be the angle between $\mathbf{u}$ and $\mathbf{v}$, ρ the angle between $\mathbf{v}$ and $\mathbf{w}$, and ϕ the angle between $\mathbf{w}$ and $\mathbf{u}$. Because $\mathbf{v} + \mathbf{w} = -\mathbf{u}$, $\theta + \frac{1}{2}\rho = 180°$. Similarly, $\rho + \frac{1}{2}\theta = 180°$, and so $\theta = \rho$ and, by symmetry, $\theta = \phi$. Therefore a, b, and c equal $re^{i\alpha}$, $re^{i(\alpha+2\pi/3)}$, and $re^{i(\alpha+4\pi/3)}$, respectively, and $a^3 = b^3 = c^3 = r^3e^{i3\alpha}$.
(c) The result is false. Simply consider ± 1 and $\pm z$ where z is on the unit circle but not equal to $\pm i$.

Note. This problem appeared in [Dod].

187. Biased Coins

It turns out that $p = (3 + \sqrt{3})/6$ yields the desired p-coin.

It is not hard to see that if p simulates $\frac{1}{2}$ and $\frac{1}{3}$, it must simulate $\frac{1}{6}$. A natural first attempt is to let $p = \frac{1}{6}$, but this doesn't work (see Problem 187.1). As a second attempt let us choose p so that if the coin is flipped twice the probability of getting HT is $\frac{1}{6}$. Such a p satisfies $p(1-p) = \frac{1}{6}$, and so p may be taken to be $(3+\sqrt{3})/6$. This p clearly simulates $\frac{1}{3}$ (consider the 2-flip event "HT or TH"; this occurs with probability $\frac{1}{6} + \frac{1}{6}$, or $\frac{1}{3}$). Now, and this is a bit of a lucky coincidence, it turns out that the 3-flip event "HHH or TTT" occurs with probability $\frac{1}{2}$; for the chance of getting one of these two sequences is

$$p^3 + (1-p)^3 = \frac{9+5\sqrt{3}}{36} + \frac{9-5\sqrt{3}}{36} = \frac{1}{2}.$$

Problem 187.1. Prove that $\frac{1}{6}$ does not simulate $\frac{1}{3}$.

Problem 187.2. Prove that $\frac{1}{6}$ does not simulate $\frac{1}{2}$.

Problem 187.3. Prove that no rational number simulates both $\frac{1}{2}$ and $\frac{1}{3}$.

Note. This problem is due to Szalkai and Velleman [SV], whose paper contains further results about coin simulations.

188. An Odd Vector Sum

Suppose $\mathbf{V}_1, \mathbf{V}_2, \ldots, \mathbf{V}_n$ are nondescending ($y \geq 0$) unit vectors. If $\mathbf{V}_1$ is neither $(1,0)$ nor $(-1,0)$, then we can decrease the length of the sum by changing it to one of these two. To prove this let $\mathbf{W} = \mathbf{V}_2 + \mathbf{V}_3 + \cdots + \mathbf{V}_n$; $\mathbf{W}$ is a nondescending vector and the sum $\mathbf{V} + \mathbf{W}$ is minimized when the angle between them is as large as possible, which requires that $\mathbf{V}$ be a horizontal unit vector. Now do the same for $\mathbf{V}_2$ (letting $\mathbf{W}$ be the sum of the new $\mathbf{V}_1$ and $\mathbf{V}_3 + \cdots + \mathbf{V}_n$), $\mathbf{V}_3$, and so on up to $\mathbf{V}_n$. This shows that given any list of rising unit vectors, we can find another list of unit vectors, all of which are horizontal, whose sum is not longer. But the sum of an odd number of horizontal unit vectors has length at least 1.

Problem 188.1. For $n \geq 2$, let U_n be the umbrella-like region that is obtained by subtracting a string of n open unit half-disks from the closed upper half-disk of radius n centered at the origin (see Figures 107–109). Show that for $n \geq 2$ the set of sums of n nondescending unit vectors coincides with U_n.

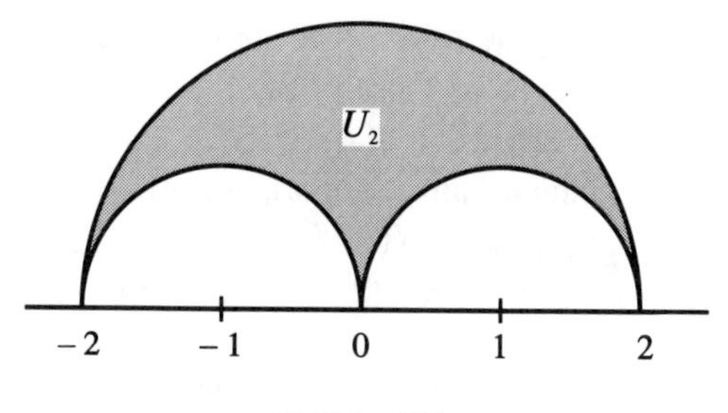

FIGURE 107

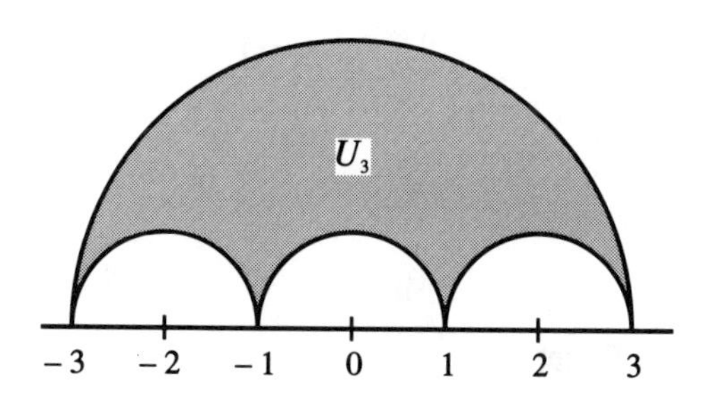

FIGURE 108

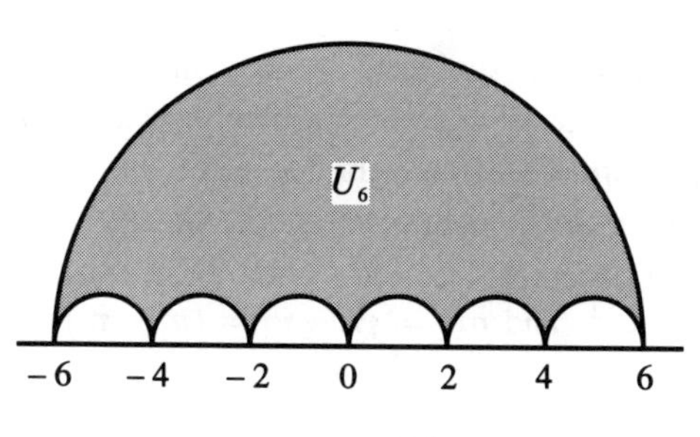

FIGURE 109

Notes. The solution to Problem 188 is due to Murray Klamkin. The problem appeared on the 1973 International Mathematical Olympiad [Gre, pp. 14–15, 148–149].

189. Positive Polynomials

Note that since $p(x) \geq 0$ for all x, n must be even and the leading coefficient of $p(x)$ must be positive.

Let $q(x) = p(x)+p'(x)+p''(x)+\cdots+p^{(n)}(x)$. Clearly the degree and leading coefficient of $q(x)$ are the same as those of $p(x)$, so $q(x)$ has a minimum value, say $q(x_0)$. It will suffice to show that $q(x_0) \geq 0$.

Since $q(x_0)$ is the minimum value of $q(x)$, $q'(x_0) = 0$. But since $p(x)$ has degree n, $p^{(n+1)}(x) = 0$, so $q'(x) = p'(x)+p''(x)+\cdots+p^{(n)}(x) = q(x)-p(x)$. Thus $q'(x_0) = q(x_0) - p(x_0) = 0$, so $q(x_0) = p(x_0) \geq 0$.

Notes. This theorem appears on p. 259 of [Hur2], and is discussed further in [Hur1]. Hurwitz gives a different proof, based on the fact that

$$\int_x^\infty \frac{p(t)}{e^t}dt = \lim_{b\to\infty}\left[-\frac{p(t)+p'(t)+\cdots+p^{(n)}(t)}{e^t}\right]_x^b$$
$$= \frac{p(x)+p'(x)+\cdots+p^{(n)}(x)}{e^x}.$$

190. Taming a Wild Function

If $n \sin n$ is to be small, then $\sin n$ must be close to 0, so n must be close to an integer multiple of π. Thus, we begin by looking for an integer multiple of π that is nearly an integer.

Let k be any positive integer, and consider the numbers $\pi, 2\pi, 3\pi, \ldots, (k+1)\pi$. For $i = 1, 2, \ldots, k+1$ write $i\pi = p_i + x_i$, where p_i is an integer and $0 < x_i < 1$. Then $x_1, x_2, \ldots, x_{k+1}$ are $k+1$ numbers between 0 and 1, so there must be two of them that differ by less than $\frac{1}{k}$. In other words, we can choose i and j such that $1 \leq i < j \leq k+1$ and $|x_j - x_i| < \frac{1}{k}$. Let $m = j-i$, $n = p_j - p_i$, and $\varepsilon = x_j - x_i$. Then $1 \leq m \leq k$, $|\varepsilon| < \frac{1}{k}$, and $m\pi = j\pi - i\pi = (p_j - p_i) + (x_j - x_i) = n + \varepsilon$. Therefore

$$|n \sin n| = |n \sin(m\pi - \varepsilon)| = n|\sin\varepsilon| \leq n|\varepsilon| < \frac{n}{k}$$
$$= \frac{m\pi - \varepsilon}{k} \leq \frac{k\pi - \varepsilon}{k} = \pi - \frac{\varepsilon}{k} < \pi + \frac{1}{k^2}.$$

Thus, provided $k \geq 2$, we have $|n \sin n| < \pi + \frac{1}{4} < 4$, as required.

To see that we can generate infinitely many different values of n in this way, note first that ε cannot be 0, since if $\varepsilon = 0$ then $\pi = \frac{n}{m}$, contradicting the irrationality of π. But since $|\varepsilon| < \frac{1}{k}$, as we increase k the procedure we have given must produce values of ε that approach 0, so it must produce infinitely many different values of ε. Since $\varepsilon = m\pi - n$ and $|\varepsilon| < 1$, ε is the difference between n and the nearest integer multiple of π. Thus different ε values must correspond to different values of n, so our procedure must produce infinitely many different values of n.

Note. It is clear from the solution we have given that the number 4 in the statement of the problem could be replaced by any number larger than π. In fact, we can do better than this. According to a theorem of Hurwitz [Niv, p. 6], for any irrational number α there are infinitely many pairs of integers (m, n) such that

$$\left|\alpha - \frac{n}{m}\right| < \frac{1}{\sqrt{5}m^2}.$$

Apply this theorem with $\alpha = \pi$, and let $\varepsilon = \pi - n/m$. Then $|\varepsilon| < 1/(\sqrt{5}m^2)$ and $n = m\pi - m\varepsilon$, so

$$\begin{aligned}
|n \sin n| &= |n \sin(m\pi - m\varepsilon)| \\
&= n|\sin m\varepsilon| \le nm|\varepsilon| < \frac{n}{\sqrt{5}m} \\
&= \frac{m\pi - m\varepsilon}{\sqrt{5}m} \\
&= \frac{\pi - \varepsilon}{\sqrt{5}} < \frac{\pi}{\sqrt{5}} + \frac{1}{5m^2}.
\end{aligned}$$

Thus, for any c $> \pi/\sqrt{5}$ there are infinitely many integers n such that $|n \sin n| < c$.

It is known that $\sqrt{5}$ is the largest number that will work in Hurwitz's theorem for all irrationals α, but in the case $\alpha = \pi$ there are larger numbers that will work. So it is possible to make $|n \sin n|$ smaller still. We do not know how small $|n \sin n|$ can be; in particular, we do not know whether or not it is true that for every $c > 0$ there are infinitely many integers n for which $|n \sin n| < c$.

191. How Many 8s Can a Plane Hold?

Suppose C is a set of disjoint figure 8s in the plane; then C must be countable. Call a point (x, y) *rational* if both x and y are rational. Because the rationals are countable, so is the set of rational points in the plane and, moreover, so is the set of pairs (P, Q) where P and Q are rational points in the plane. Recall also that the rationals are dense in $\mathbf{R}$, which implies that the rational points are dense in the plane; that is, any disk in the plane contains a rational point. Now, to each figure 8 in C associate a pair (P, Q) by choosing P inside one loop of the 8 and Q inside the other. The interior of a loop contains a disk, and so such rational points can be found. Because the figure 8s cannot cross each other, two figure 8s cannot be assigned the same pair (P, Q). In other words, this assignment of pairs is a one-to-one correspondence of C with a subset of a countable set. So C is itself countable. The same proof applies if the 8s can be deformed continuously.

Problem 191.1. What if the 8s are replaced by 4s, where a four is written as 4? What about other digits, written in various ways?

Notes. The result of Problem 191 can be extended to: A graph with uncountably many vertices of degree 3 or more is not planar. This is an old result of R. L. Moore [Moo]. Moreover, such a graph is essentially the only counterexample to the assertion that the surface on which a graph embeds is directly related to the surfaces on which its finite subgraphs embed. For more precision and proofs regarding this statement see [HW].

References

[AB] Shmuel Avital and Ed Barbeau, Nonintuitive solutions to problems, *For the Learning of Mathematics* 11:3 (1991) 2–18.

[Abb] Paul Abbott, Tricks of the Trade: Recursively defined function, *The Mathematica Journal* 5:2 (1995) 58–59.

[AC] Michael Albertson and Karen Collins, Symmetry breaking in graphs, *Electronic Journal of Combinatorics* 3 (1996).

[Ahl] Hayo Ahlburg, Problem 639, *Crux Mathematicorum* 7 (1981) 146; 8 (1982) 145–148.

[AKW] J. Aldens, J. S. Kline, and S. Wagon, Problem 633, *Crux Mathematicorum* 7 (1981) 145; 8 (1982) 120–121.

[Ale] G. L. Alexanderson, A conversation with Leon Bankoff, *The College Mathematics Journal* 23 (1992) 98–117.

[Ann] Norman Anning, Problem 203, *Mathematics Magazine* 27 (1954) 279; 28 (1955) 163.

[Ban] Leon Bankoff, Regular polygons of 2, 3, 4 and 6 sides inscribed in circles of unit radius, *Scripta Mathematica* 21 (1955) 252.

[BCCG] Robert C. Brigham, Richard M. Caron, Phyllis Z. Chinn, and Ralph Grimaldi, A tiling scheme for the Fibonacci numbers, *Journal of Recreational Mathematics* (to appear).

[BCHW] Robert C. Brigham, Phyllis Z. Chinn, Linda Holt, and Steve Wilson, Finding the recurrence relation for tiling $2 \times n$ rectangles, Twenty-fifth Southeastern International Conference on Combinatorics, Graph Theory and Computing, Boca Raton, Florida, March 7–11, 1994, *Congressus Numerantium* 105 (1994) 134–138.

[Ber] G. Berzsenyi, Problems, puzzles, and paradoxes, *Consortium,* Number 16, winter, 1985, 4.

[Ber1] ——, Problems, puzzles, and paradoxes, *Consortium,* fall, 1992, 12.

[BK] Alfred Brauer and Aubrey Kempner, Problem E1555, *American Mathematical Monthly* 69 (1962) 1009; 70 (1963) 896–897.

[BLIC] The Beardstown Ladies Investment Club with Leslie Whitaker, *The Beardstown Ladies' Common-Sense Investment Guide,* Hyperion, New York, 1994.

[BM] J. A. Bondy and U. S. R. Murty, *Graph Theory with Applications,* North-Holland, New York, 1978.

[BO] M. D. Beeler and P. E. O'Neil, Some new van der Waerden numbers, *Discrete Mathematics* 28 (1979) 135–146.

[Bol] Brian Bolt, *Mathematics Meets Technology,* Cambridge University Press, Cambridge, 1991.

[Bot] O. Bottema, Problem S34, *American Mathematical Monthly* 87 (1980) 574; 89 (1982) 592–593.

[Cam] W. B. Campbell, Problem E106, *American Mathematical Monthly* 41 (1934) 447; 42 (1935) 110–111.

[CCFM] Phyllis Z. Chinn, Greg Colyer, Martin Flashman, and Ed Migliore, Cuisenaire rods go to college, *Primus* 2 (1992) 118–130.

[CDW] J. Duncan, L. Carter, and S. Wagon, Problem 1457, *Mathematics Magazine* 67 (1994) 304; 68 (1995) 312–315.

[CFG] Hallard T. Croft, Kenneth J. Falconer, and Richard K. Guy, *Unsolved Problems in Geometry,* Springer-Verlag, New York, 1991.

[CLL] Jesse Chan, Peter Laffin, and Da Li, Martin Gardner's "Royal Problem," *Quantum* 4:1 (1993) 45–46.

[Con] J. H. Conway, *On Numbers and Games,* London Mathematical Society Monographs No. 6, Academic Press, London, 1976.

[Cox] H. S. M. Coxeter, *Introduction to Geometry,* 2nd edition, Wiley, New York, 1969.

[Cro] Hallard T. Croft, 9-point and 7-point configurations in 3-space, *Proceedings of the London Mathematical Society* 12 (1962) 400–424.

[Cur] D. R. Curtiss, On Kellogg's Diophantine problem, *American Mathematical Monthly* 29 (1922) 380–387.

[CV] Gregory Call and Dan Velleman, Permutations and combination locks, *Mathematics Magazine* 68 (1995) 243–253.

[CW] L. Carter and S. Wagon, Proof Without Words: Fair allocation of a pizza, *Mathematics Magazine* 67 (1994) 267.

[DDS] Joseph DiStefano, Peter Doyle, and Laurie Snell, The evil twin strategy for a football pool, *American Mathematical Monthly* 100 (1993) 341–343.

[Dic] L. E. Dickson, *History of the Theory of Numbers,* vol. II, Chelsea, Bronx, 1971.

[Dod] Clayton Dodge, Problem 353, *Pi Mu Epsilon Journal* 6 (1975) 178, 313.

[Doo] Michael Doob, *The Canadian Mathematical Olympiad, 1969–1993,* Canadian Mathematical Society, Ottawa, 1993.

[Dor] Heinrich Dörrie, *100 Great Problems of Elementary Mathematics,* Dover, New York, 1965.

[dSV] Ricardo E. de Souza and Giovani L. Vasconcelos, Visual appearance of a rolling bicycle wheel, *American Journal of Physics* 64 (1996) 896–897.

[Dud] H. E. Dudeney, *536 Puzzles and Curious Problems,* ed. Martin Gardner, Scribner's, New York, 1967.

[DVW] J. Duncan, D. Velleman, and S. Wagon, Problem 2006, *Crux Mathematicorum* 21 (1995) 20; 22 (1996) 37–38.

[Ell] D. Elliot, M. L. Urquhart, *Journal of the Australian Mathematical Society* 8 (1968) 129–133.

[Eng] Arthur Engel, Geometrical activities for the upper elementary school, *Educational Studies in Mathematics* 3 (1971) 353–394.

[Erd1] P. Erdős, Egy Kürschák-féle elemi számelméleti tétel általánosítása, *Matematikai és Fizikai Lapok* 39 (1932) 1–8.

[Erd2] ——, Problem E735, *American Mathematical Monthly* 53 (1946) 394; 54 (1947) 227–229.

[ES] S. J. Einhorn and I. J. Schoenberg, On Euclidean sets having only two distances between points. I and II, *Indagationes Mathematicae* 28 (1966) 479–504.

[ESz] Paul Erdős and George Szekeres, Some number theoretic problems on binomial coefficients, *Australian Mathematical Society Gazette* 5 (1978) 97–99.

[Eus] Dan Eustice, Urquhart's theorem and the ellipse, *Crux Mathematicorum* (Eureka) 2 (1976) 132–133.

[Fen] Daniel Fendel, Prime-producing polynomials and principal ideal domains, *Mathematics Magazine* 58 (1985) 204–210.

[Fla] Daniel Flath, *Introduction to Number Theory,* Wiley, New York, 1989.

[FP] H. Fukagawa and D. Pedoe, *Japanese Temple Geometry Problems,* The Charles Babbage Research Center, Winnipeg, 1989.

[FR] P. C. Fishburn and J. A. Reed, Unit distances between vertices of a convex polygon, *Computational Geometry: Theory and Applications* 2 (1992) 81–91.

[Fri] Gerd Dricke, Sandra Hedetniemi, Stephen Hedetniemi, Alice McRae, Charles Wallis, Michael Jacobson, Harold Martin, and William Weakley, Combinatorial problems on chessboards: A brief survey, in *Graph Theory, Combinatorics, and Applications, Proceedings of the Seventh Quadrennial International Conference on the Theory and Applications of Graphs,* Western Michigan University, Y. Alavi and A. Schwenk, eds., Wiley, New York, 1995, 507–528.

[Fuk] J. Fukuta, Problem 1426, *Mathematics Magazine* 66 (1993) 192; 67 (1994) 227–228.

[Gar] Jack Garfunkel, Problem 553, *Pi Mu Epsilon Journal* 7 (1983) 614; 8 (1984) 57.

[Gar1] Martin Gardner, *Penrose Tiles to Trapdoor Ciphers,* W. H. Freeman, New York, 1989.

[Gar2] ——, *Fractal Music, Hypercards, and More,* W. H. Freeman, New York, 1992.

[Gil] C. Gill, *Application of the Angular Analysis to Indeterminate Problems of the Second Degree,* Wiley, New York, 1848.

[GKP] Ronald L. Graham, Donald E. Knuth, and Oren Patashnik, *Concrete Mathematics,* Addison-Wesley, Reading, Mass., 1989.

[GL] Martin Gardner and Andy Liu, A royal problem, *Quantum* 3:6 (1993) 30–31.

[Gob] F. Göbel, Geometrical packing and covering problems, in *Packing and Covering in Combinatorics,* A. Schrijver, ed., Tweede Boerhaavestraat, Amsterdam, 1979.

[Gol] M. Goldberg, On the original Malfatti problem, *Mathematics Magazine* 40 (1967) 241–247.

[Gol1] Alexander S. Golovanov, The Malfatti problem is solved, *American Mathematical Monthly,* to appear.

[Gre] Samuel L. Greitzer, *International Mathematical Olympiads 1959–1977,* New Mathematical Library, vol. 27, Mathematical Association of America, Washington, D.C., 1978.

[Guy] Richard K. Guy, *Unsolved Problems in Number Theory,* Springer, New York, 1981.

[Har1] Heiko Harborth, Match sticks in the plane, in *The Lighter Side of Math, Proceedings of the Eugène Strens Memorial Conference on Recreational Mathematics and its History,* Mathematical Association of America, Washington, D.C., 1994, 281–288.

[Har2] ——, Regular point sets with unit distances, *Colloquia Mathematica Societatis János Bolyai* 48 (1985) 239–253.

[HK] H. S. Hall and S. R. Knight, *Higher Algebra,* 3rd ed., Macmillan, London, 1890.

[HK1] Heiko Harborth and Arnfried Kemnitz, Integral representation of graphs, in *Contemporary Methods in Graph Theory,* R. Bodendiek and R. Henn, eds., BI-Wiss.-Verl., Mannheim, Germany, 1990, 359–367.

[Hon] Ross Honsberger, *Mathematical Gems II,* Dolciani Mathematical Expositions, no. 2, Mathematical Association of America, Washington, D.C., 1976.

[Hon1] Ross Honsberger, *From Erdős to Kiev: Problems of Olympiad Caliber,* Dolciani Mathematical Expositions, no. 17, Mathematical Association of America, Washington, 1996.

[HP] Heiko Harborth and Lothar Piepmeyer, On special integral Erdős point sets, *Colloquia Mathematica Societatis János Bolyai* (to appear).

[HR] Nora Hartsfield and Gerhard Ringel, *Pearls in Graph Theory: A Comprehensive Introduction,* Academic Press, New York, 1990.

[Hur1] A. Hurwitz, Über definite polynome, *Mathematische Annalen* 73 (1913) 173–176.

[Hur2] ——, Über die nullstellen der Bessel'schen function, *Mathematische Annalen* 33 (1889) 246–266.

[HW] Joan P. Hutchinson and Stan Wagon, A forbidden subgraph characterization of infinite graphs having finite genus, in *Graphs and Applications, Proceedings of the First Colorado Symposium on Graph Theory,* F. Harary, J. S. Maybee, eds., Wiley, New York, 1985, 183–194.

[Jac] Brad Jackson, Universal cycles of k-subsets and k-permutations, *Discrete Mathematics* 177 (1993) 141–150.

[JD] Karl-Georg Jacobson and Andrejs Dunkels, Problem 1071, *Journal of Recreational Mathematics* 14 (1982) 140; 15 (1983) 149.

[Kat] Zelda Katz, Problem 494, *Pi Mu Epsilon Journal* 7 (1981) 265.

[Ken] Richard Kenyon, A note on tiling with integer-sided rectangles, *Journal of Combinatorial Theory A,* to appear.

[KL] M. S. Klamkin and A. Liu, Three more proofs of Routh's theorem, *Crux Mathematicorum* 7 (1981) 199–203.

[Kla1] Murray Klamkin, The teaching of mathematics so as to be useful, *Educational Studies in Mathematics* 1 (1968) 126–160.

[Kla2] ——, *USA Mathematical Olympiads 1972–1986,* New Mathematical Library, vol. 33, Mathematical Association of America, Washington, D.C., 1988.

[Kle1] V. Klee, The generation of affine hulls, *Acta Scientarum Mathematicorum (Szeged)* 24 (1963) 60–81.

[Kle2] ——, Problem 1413, *Mathematics Magazine* 66 (1993) 56; 67 (1994) 68–69.

[Kon] Joseph D. E. Konhauser, Puzzle 3, *Pi Mu Epsilon Journal* 8 (1988) 595.

[Kri] Mary S. Krimmel, Problem 3978, *School Science and Mathematics* 84 (1984) 84, 713.

[KV] James S. Kline and Dan Velleman, Yet another proof of Routh's theorem, *Crux Mathematicorum* 21 (1995) 37–40.

[KW] Victor Klee and Stan Wagon, *Old and New Unsolved Problems in Plane Geometry and Number Theory,* Dolciani Mathematical Expositions No. 11, Mathematical Association of America, Washington, D.C., 1991.

[Lar] Loren C. Larson, *Problem-Solving Through Problems,* Springer-Verlag, New York, 1983.

[Lar1] ——, Problem 769, *Crux Mathematicorum* 8 (1982) 210; 9 (1983) 284–285.

[Lew] Leonard Lewin, *Polylogarithms and Associated Functions,* Elsevier North Holland, New York, 1981.

[Lin] Viktors Linis, Problem 232, *Crux Mathematicorum* 3 (1977) 104; 3 (1977) 238–240; 4 (1978) 17–18.

[Liu1] Andy Liu, Problem 83-11, *The Mathematical Intelligencer* 6:4 (1984) 39, 43.

[Liu2] ——, On the "Seven Points Problem" of P. Erdős, *Mathematics Chronicles* 15 (1986) 29–33.

[Liu3] ——, Problem 92-2, *SIAM Review* 35 (1993) 137–140.

[Lub] D. Lubell, A short proof of Sperner's lemma, *Journal of Combinatorial Theory* 1 (1966) 299.

[Mak] Andrzej Makowski, Comment on problem 533, *Mathematics Magazine* 42 (1969) 275–276.

[Mar] J. Marica, On a conjecture of Conway, *Canadian Mathematical Bulletin* 12 (1969) 233–234.

[Mau] J. G. Mauldon, Problem E3023, *American Mathematical Monthly* 90 (1983) 645; 96 (1989) 254–258.

[McC] Edwin P. McCravy, Problem 940, *Mathematics Magazine* 48 (1975) 180; 49 (1976) 152–153.

[Moo] R. L. Moore, Concerning triods in the plane and the junction points of plane continua, *Proceedings of the National Academy of Sciences of the USA* 14 (1928) 85–88.

[Nel] Harry L. Nelson, Letter to the editor, *Journal of Recreational Mathematics* 20 (1988) 315–316.

[Niv] Ivan Niven, *Diophantine Approximations,* Interscience, New York, 1963.

[NZM] Ivan Niven, Herbert S. Zuckerman, and Hugh L. Montgomery, *An Introduction to the Theory of Numbers,* 5th ed., Wiley, New York, 1991.

[Odo] George Odom, Problem E3007, *American Mathematical Monthly* 90 (1983) 482; 93 (1986) 572.

[Pal] Ilona Palásti, Lattice point examples for a question of Erdős, *Periodica Mathematica Hungarica* 20 (1989) 231–235.

[Ped] Dan Pedoe, The most "elementary" theorem of Euclidean geometry, *Mathematics Magazine* 49 (1976) 40–42; also letters, ibid., 49 (1976) 261 and 50 (1977) 55.

[Pic] Sophie Piccard, Sur des ensembles parfaits, *Mémoires de l'université de Neuchatel* 16 (1942).

[Pie] J. L. Pietenpol, Problem E1571, *American Mathematical Monthly* 70 (1963) 330; 71 (1964) 91–92.

[Pos] K. A. Post, Letter, *Journal of Recreational Mathematics* 11 (1978–1979) 41.

[Rab1] Stanley Rabinowitz, Problem 3936, *School Science and Mathematics* 83 (1983) 172; 84 (1984) 86.

[Rab2] ——, Problem 535, *Pi Mu Epsilon Journal* 7 (1983) 542; 7 (1984) 676–677.

[Rab3] ——, Problem 660, *Pi Mu Epsilon Journal* 8 (1987) 470; 9 (1988) 617.

[Rab4] ——, Problem 1325, *Crux Mathematicorum* 14 (1988) 77; 15 (1989) 120–122.

[Rab5] ——, *Ptolemy's Legacy,* MathPro Press (to appear).

[Ran] Ernest R. Ranucci, Problem 156, *Journal of Recreational Mathematics* 4 (1971) 67; 5 (1972) 151–152.

[Ros] Moshe Rosenfeld, A dynamic puzzle, *American Mathematical Monthly* 98 (1991) 22–24.

[Rub] Frank Rubin, Problem 729, *Journal of Recreational Mathematics* 11 (1979) 128; 12 (1980) 145.

[Ruz] Imre Ruzsa, Sets of sums and differences, *Séminaire de Théorie des nombres de Paris 1982–83,* Birkhauser-Boston (1984), 267–273.

[Sal1] Lee Sallows, A new type of crossword puzzle, *American Mathematical Monthly* 94 (1987) 666.

[Sal2] ——, New pathways in serial isogons, *The Mathematical Intelligencer* 14:2 (1992) 55–67.

[Sal3] ——, *Word Ways* (to appear).

[SB] Sherman K. Stein and Anthony Barcellos, *Calculus and Analytic Geoemtry,* 5th ed., McGraw-Hill, New York, 1992.

[Sch] Don Scholten, Problem 13, Flaring the circle, *The Old Farmer's Almanac,* no. 196 (1988), 199.

[SCY] D. O. Shklarsky, N. N. Chentzov, and I. M. Yaglom, *The USSR Olympiad Problem Book,* revised and edited by I. Sussman, trans. by J. Maykovich, W. H. Freeman, San Francisco, 1962.

[SGGK] Lee Sallows, Martin Gardner, Richard Guy, Donald Knuth, Serial isogons of 90 degrees, *Mathematics Magazine* 64 (1991) 315–324.

[Sie] Waclaw Sierpiński, *250 Problems in Elementary Number Theory,* American Elsevier, New York, 1970.

[Sil1] David Silverman, Problem 533, *Mathematics Magazine* 36 (1963) 319; 37 (1964) 201–202.

[Sil2] ——, Problem 71, *Journal of Recreational Mathematics* 2 (1969) 99.

[Sil3] ——, Problem 73, *Journal of Recreational Mathematics* 2 (1969) 100; 3 (1970) 51.

[Skv] Ivan Skvarca, Problem 1752, *Journal of Recreational Mathematics* 21 (1989) 317; 22 (1990) 304.

[Smi] Leander W. Smith, Conditions governing numerical equality of perimeter, area, and volume, *The Mathematics Teacher* 58 (1965) 303–307.

[Soi] Alexander Soifer, *How Does One Cut a Triangle?,* Center for Excellence in Mathematical Education, Colorado Springs, Colo., 1990.

[Sok1] Dan Sokolowsky, Extensions of two theorems of Grossman, *Crux Mathematicorum* (Eureka) 2 (1976) 163–170.

[Sok2] ——, A "no-circle" proof of Urquhart's theorem, *Crux Mathematicorum* (Eureka) 2 (1976) 133–134.

[Spe] E. S. Sperner, Ein Satz über Untermengen einer endlichen Menge, *Mathematische Zeitschrift* 27 (1928) 544–548.

[Spr] R. Sprague, Über Zerlegungen in n-te Potenzen mit lauter vershiedenen Grundzahlen, *Mathematische Zeitschrift* 51 (1949) 466–468.

[Ste] Sherman K. Stein, On the cardinalities of $A + A$ and $A - A$, *Canadian Mathematical Bulletin* 16 (1973) 343–345.

[Ste1] Hugo Steinhaus, *Mathematical Snapshots,* revised and enlarged edition, Oxford University Press, New York, 1960.

[Ste2] ——, *One Hundred Problems in Elementary Mathematics,* Pergamon Press, London, 1963.

[Stew] Ian Stewart, *Another Fine Math You've Got Me Into,* W. H. Freeman, New York, 1992.

[Str] S. Straszewicz, *Problems and Puzzles from the Polish Mathematical Olympiads,* trans. J. Smólska, Popular Lectures in Mathematics, vol. 12, Pergamon Press, Oxford, 1965.

[SV] István Szalkai and Dan Velleman, Versatile coins, *American Mathematical Monthly* 100 (1993) 26–33.

[Sza] Sándor Szabó, On finite abelian groups and parallel edges on polygons, *Mathematics Magazine* 66 (1993) 36–39.

[Sze] G. Szekeres, Kinematic geometry, M. L. Urquhart in memoriam, *Journal of the Australian Mathematical Society* 8 (1968) 134–160.

[Szi] T. Szirtes, On the problem of the interchangeable clock hands, *Journal of Recreational Mathematics* 8 (1975–1976) 159–168.

[Tay] P. J. Taylor, *International Mathematics Tournament of the Towns, 1980 to 1984, Questions and Solutions,* Australian Mathematics Trust, Belconnen, Australia, 1993, 867–879.

[Tok] George Tokarsky, Polygonal rooms not illuminable from every point, *American Mathematical Monthly* 102 (1995) 867–879.

[Tri1] Charles W. Trigg, Problem 236, *Mathematics Magazine* 28 (1955) 283; 29 (1956) 164.

[Tri2] ——, Integers immune to partitioning into distinct squares, *Journal of Recreational Mathematics* 3 (1970) 124.

[Tri3] ——, Absolute difference triangles, *Journal of Recreational Mathematics* 9 (1976–1977) 271–75.

[Uma] Harlan L. Umansky, Problem 644, *Mathematics Magazine* 40 (1967) 42; 224–225.

[Upt] L. J. Upton, Problem 660, *Mathematics Magazine* 40 (1967) 163; 41 (1968) 46.

[Wag1] Stan Wagon, Problem 83-9, *The Mathematical Intelligencer* 5:3 (1983) 45–46.

[Wag2] ——, Fourteen proofs of a result about tiling a rectangle, *American Mathematical Monthly* 94 (1987) 601–17. Reprinted in *The Lighter Side of Mathematics, Proceedings of the Eugène Strens Memorial Conference on Recreational Mathematics and its History,* Mathematical Association of America, Washington, D.C., 1994, 113–128.

[Wag3] ——, Problem 799, *Pi Mu Epsilon Journal* 9 (1993) 544; 9 (1994) 694–696.

[Wag4] ——, Quintuples with square triplets, *Mathematics of Computation* 64 (1995) 1755–1756.

[Wag5] ——, Programming tips: Bézier curves, *Mathematica in Education* 4:1 (1995) 48–53.

[Way] Alan Wayne, Problem 411, *Crux Mathematicorum* 5 (1979) 46–47; 5 (1979) 299–300.

[Wei] Gerald Weinstein, Problem 10260, *American Mathematical Monthly* 99 (1992) 873; 102 (1995) 364.

[Wie] Norbert Wiener, The shortest line dividing an area in a given ratio, *Proceedings of the Cambridge Philosophical Society* 18 (1914) 56–58.

[Wil] Kenneth S. Williams, On Urquhart's elementary theorem of Euclidean geometry, *Crux Mathematicorum* (Eureka) 2 (1976) 108–109.

[Wil1] Kenneth Wilke, Problem 455, *Pi Mu Epsilon Journal* 7 (1979) 58; 7 (1980) 196.

[Wil2] Kenneth Wilke, Problem 375, *The Pentagon* 43 (1984) 113; 44 (1985) 120–124.

[Wils] David C. Wilson, Problem 3908, *School Science and Mathematics* 82 (1982) 440; 83 (1983) 358.

[WW] Stan Wagon and Herbert S. Wilf, When are subset sums equidistributed modulo m?, *Electronic Journal of Combinatorics* 1 (1994).

[WWo] Stan Wagon, Olympiad Corner (ed. R. Woodrow), *Crux Mathematicorum* 19 (1993) 260–261; 20 (1994) 45.

[Zai] Zhang Zaiming, Problem 474, *College Mathematics Journal* 23 (1992) 163; 24 (1993) 188.

[ZL] V. A. Zalgaller and G. A. Loš, Solution of the Malfatti optimization problem, *Ukr. Geom. Sbornik* 35 (1992) 14–33 (Russian).

INDEX

Joseph D. E. Konhauser was an avid problemist throughout his years at Macalester College (1968–1991). He studied at Penn State University (bachelor's, 1948; doctorate, 1963) and held teaching positions at Penn State and the University of Minnesota before coming to Macalester. Joe was a very active problemist and served on many contest committees such as those governing the USA Mathematical Olympiad and the William Lowell Putnam Mathematics Competition. He also enjoyed editorial work; he edited the *Pi Mu Epilson Journal* and served as book review editor for the *American Mathematical Monthly*.

Dan Velleman received his bachelor's degree from Dartmouth College in 1976 and his doctorate from the University of Wisconsin in 1980. He has taught at the University of Texas and the University of Toronto, and since 1983 he has taught at Amherst College. In addition to enjoying mathematical problems, he is interested in logic, the philosophy of mathematics, and the foundations of quantum mechanics. He is the author of *How to Prove It* (Cambridge University Press).

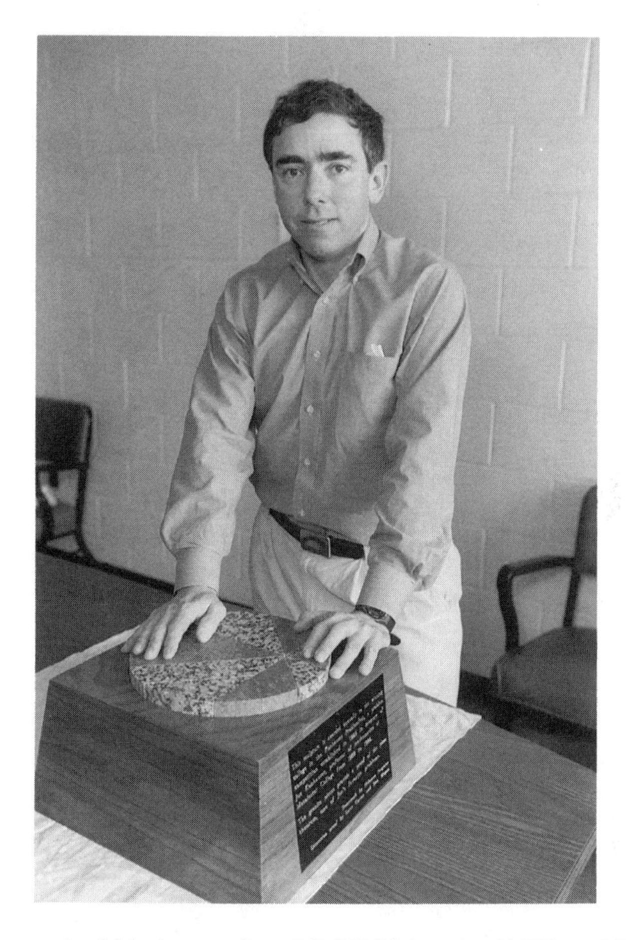

Stan Wagon received his degrees from McGill University (1971) and Dartmouth College (1975). He taught at Smith College until coming to Macalester in 1990. Throughout his career he has enjoyed the special beauty of succinctly stated and surprising mathematical facts. This led to his book on the Banach–Tarski paradox (Cambridge University Press) and, with Victor Klee, a book on unsolved problems in elementary mathematics (MAA). Recently he has been intrigued with how *Mathematica* can help us see mathematical constructions in new ways, and he has written several books illustrating the power of this software: *Mathematica in Action* (W. H. Freeman), *The Power of Visualization* (Front Range Press), *Animating Calculus* (Springer/TELOS; with Ed Packel), and *VisualDSolve: Visualizing Ordinary Differential Equations with Mathematica* (Springer/TELOS; with Dan Schwalbe).

The granite sculpture illustrates a solution to the following intriguing problem of elementary geometry: Suppose four cuts are made through a point P in a disk so that the eight angles at P are all equal to 45°. If the resulting eight pizza slices are colored alternately black and white, must the white area equal the black area?

The granite pizza, designed by Helaman Ferguson, illustrates a dissection solution to this problem. For a fuller discussion, see problems 62 and 63 of this book.